SWITCHED RELUCTANCE MOTOR - CONCEPT, CONTROL AND APPLICATIONS

Edited by **Ahmed Tahour**
and **Abdel Ghani Aissaoui**

Switched Reluctance Motor - Concept, Control and Applications
http://dx.doi.org/10.5772/66849
Edited by Ahmed Tahour and Abdel Ghani Aissaoui

Contributors

Christopher H. T. Lee, Ebrahim Amiri, Mohammad Mahdi Bouiabady, Aliakbar Damaki Aliabad, Sandeep Narla, Erdal Sehirli, Meral Altınay, Mircea Ruba, Petre Dorel Teodosescu, Srinivas Pratapgiri

Published by InTech

Janeza Trdine 9, 51000 Rijeka, Croatia

Publishing Process Manager Dajana Pemac
Technical Editor SPi Global
Cover InTech Design team

First published June, 2017
Printed in Croatia
Legal deposit, Croatia: National and University Library in Zagreb

Additional hard copies can be obtained from orders@intechopen.com

Switched Reluctance Motor - Concept, Control and Applications, Edited by Ahmed Tahour and Abdel Ghani Aissaoui
p. cm.
Print ISBN 978-953-51-3267-7
Online ISBN 978-953-51-3268-4

For EU product safety concerns:
IN TECH d.o.o., Prolaz Marije Krucifikse Kozulić 3, 51000 Rijeka, Croatia,
info@intechopen.com or visit our website at intechopen.com.

Meet the editors

Ahmed Tahour was born in 1972 in Ouled Mimoun, Tlemcen, Algeria. He received his BS degree in electrical engineering in 1996, the MS degree in 1999 and the PhD in 2007 from the Electrical Engineering Institute of the University of Sidi Bel Abbes (Algeria). He is currently a full Professor of electrical engineering at University of Mascara (Algeria). His current research interest includes power electronics, control of electrical machines and renewable energies.

Abdel Ghani Aissaoui is a full Professor of electrical engineering at the University of Bechar (Algeria). He was born in 1969 in Naama, Algeria. He received his BS degree in 1993, the MS degree in 1997 and the PhD degree in 2007 from the Electrical Engineering Institute of Djilali Liabes University of Sidi Bel Abbes (Algeria). He is an active member of IRECOM (Interaction Réseaux Electriques – COnvertisseurs Machines) Laboratory and IEEE senior member. He is an editor member and a reviewer in many international journals. He serves as a member in technical committee (TPC) and as a reviewer in international conferences (CHUSER 2011, SHUSER 2012, PECON 2012, SAI 2013, SCSE 2013, SDM 2014, SEB 2014, PEMC 2014, PEAM 2014, SEB [2014, 2015], ICRERA 2015, etc.). His current research interest includes power electronics, control of electrical machines, artificial intelligence and renewable energies.

Contents

Preface IX

Chapter 1 **Switched Reluctance Motor Topologies: A Comprehensive Review 1**
Mohammad Mahdi Bouiabady, Aliakbar Damaki Aliabad and Ebrahim Amiri

Chapter 2 **Four-Quadrant Control of Switched Reluctance Machine 25**
Sandeep Narla

Chapter 3 **Direct Instantaneous Torque Controlled Switched Reluctance Motor Drive for Fan Type Load and Constant Torque Load 43**
Srinivas Pratapgiri

Chapter 4 **Design, Power Electronics and Torque Control of Switched Reluctance Machines 59**
Mircea Ruba and Petre Dorel Teodosescu

Chapter 5 **Current-Controlled SRM Fed by Three-Phase Boost PFC 79**
Erdal Şehirli and Meral Altınay

Chapter 6 **Switched Reluctance Motor Drives for Hybrid Electric Vehicles 97**
Christopher H.T. Lee, James L. Kirtley, Jr. and M. Angle

Preface

During the last two decades, remarkable technological developments have been made in the field of power electronics and computing. This resumed works on the switched reluctance machines (SRMs). The new computer-aided conception techniques have made it possible to describe different high performances of SRM structures. In addition, new high-performance power switches and new converter topologies have enabled an improved power output of the motor allowing a much more efficient operation of the machine. At the same time, works were undertaken on both modeling and control.

Because of its electromagnetic structure, the switched reluctance (SR) motor develops a strongly pulsating torque. The stator of the switched reluctance motor has a set of concentrated excitation coils, and the rotor has no winding; the two components of the machine are made of soft iron steel. The production of torque is made when two opposite poles are excited. Moreover, due to its operation in saturation and its variable reluctance, the SR motor is a system whose model is strongly nonlinear and uncertain.

The goal of this book is to present recent works on concept, control, and applications in switched reluctance machine. The developments of this field go fast, accompanied with many difficulties that required solutions. Different solutions are proposed based on new techniques of control and advanced technology products in the field of switched reluctance motors.

The book is intended for researchers in the field of design and control of electrical machines. For anyone working in the development of reluctance motor drive systems, it is hoped that it will serve as useful reference work.

This book is divided as follows:

- In the first chapter, authors present a comprehensive technology status review highlighting advantages and drawbacks of each topology as well as recent trends in switched reluctance machines in incorporating the permanent magnets (PMs) in the motor design.
- In the second chapter, the control of the switches inside the converter manages continuous operation from each phase of the SRM; the turn-on and turn-off angles of the operation dictate the mode of operation such as forward motoring, reverse motoring, forward generating, and reverse generating.
- A complete analytic design, different electronic converters, and detailed control strategies are presented in the third chapter. The results are validated via simulations and experimental measurements.

- The fourth chapter describes a comparative analysis of direct instantaneous torque control (DITC)-based drive with constant torque load and fan-type load.
- The fifth chapter deals with converter topologies for SRM. Different converter topologies are introduced. Then, SRM is fed by nonlinear controlled three-phase boost power factor correction (PFC) in order to provide energy efficiency. Converter that feeds SRM includes uncontrolled three-phase rectifier and DC-DC boost converter. DC-DC boost converter is operated by high switching frequency and is connected after three-phase rectifier.
- In the sixth chapter, authors present an overview of the switched reluctance motor drives (SRMDs) for electric vehicle (EV) and hybrid electric vehicle (HEV) applications. The quantitative comparisons among the SRMDs and their major counterparts are made, while the current and upcoming technologies are also covered. All the covered materials will serve as a blueprint and start-up manual for the potential readers to develop their own interests in this particular area, i.e., to develop the new research directions of SRMDs for EV and HEV systems.

The chapters of this book present recent works in the concept, control, and applications in switched reluctance machines. Contributions reveal the importance of the treated subject. We aim by this book to give new trends of the researches in electrical machine areas and to present the new challenges in the implementation of new construction of switched reluctance machine. We hope that the readers will find this book a unique and significant source of knowledge and reference for the future years.

Prof. Dr. Ahmed Tahour
Electrical Department, Faculty of Technology
University of Mascara, Algeria

Prof. Dr. Abdel Ghani Aissaoui
Electrical Department, Faculty of Technology
University Tahri Mohamed of Bechar (UTMB), Algeria

Chapter 1

Switched Reluctance Motor Topologies: A Comprehensive Review

Mohammad Mahdi Bouiabady,
Aliakbar Damaki Aliabad and Ebrahim Amiri

Additional information is available at the end of the chapter

http://dx.doi.org/10.5772/intechopen.69149

Abstract

Switched reluctance motor (SRM) is gaining much interest in industrial applications such as wind energy systems and electric vehicles due to its simple and rugged construction, high-speed operation ability, insensitivity to high temperature, and its features of fault tolerance. With continued research, different topologies have emerged presenting less torque ripple, high efficiency, high power factor, and high power density. However, there has always been a trade-off between gaining some of the advantageous and losing some with each new technology. In this chapter, various SRM topologies, design, principle of operation, and respective phase switching schemes are extensively reviewed, and their advantages and drawbacks are discussed. On the other hand, some of SRM limitations (such as excitation penalty, control complexity, noise, and vibration) have prompted research into the incorporation of permanent magnets into the basic SRM structure, and therefore, the chapter also includes discussion on a new class of SRM with permanent magnet assist (PM-assist) called doubly salient permanent magnet (DSPMM). The DSPM motor incorporates the merits of both the PM brushless motor and the SRM.

Keywords: comprehensive review, switched reluctance motors, doubly salient permanent magnet, E-core structure

1. Introduction

The original idea of switched reluctance motors (SRMs) dates back to 1814; however, these motors were reinvented and came into practical use in recent decades in line with the development of power electronic devices. Switched reluctance motors have salient poles in both the rotor and the stator and act as a single-excited configuration with inactive (coil-free) rotors. The stator has a centralized winding system with multiple phases. The coils are fed regularly and sequentially from a DC power supply, and thus, they generate electromagnetic torque. Because of their simplicity and structural strength, SRMs have been of great interest in the past two decades, and they are expected to find broader applications in terms of price and quality compared to other motors. In addition, many studies have been carried out to enhance the performance of these motors as potential alternative to AC (asynchronous and synchronous) motors. At present, switched reluctance motors are in their infancy in commercial terms, but it is expected that they will be used more widely in the near future [1].

SRMs can be considered a stepping motor in type. However, important differences in their configuration and methods used to control them have placed switched reluctance motors into a separate category. The most important differences between SRMs and variable stepper reluctance motors are as follows:

- SRMs have much bigger steps but much fewer poles than steppers.
- SRMs have a closed-loop control system while steppers have specific steps and operate without the use of feedback and in an open loop.

Like other reluctance motors, SRMs usually have no magnets in the rotor and stator, and thus, they enjoy a simple, cheap, and firm structure. However, a small amount of permanent magnetic materials is used in some types of these motors to improve their torque. The number of the ratio of stator poles to rotor poles used in SRMs is rather limited, the most common of which are 4:6 and 6:8, each with its own possible multipliers. It should be noted that each pair of coils facing each other makes up a phase. Therefore, for 4:6 and 6:8 ratios, there will be three phases and four phases on the stator, respectively. **Figure 1** depicts a three-phase and four-phase SRM [2, 3].

In two-phase structures, the problem of nongeneration of the starting torque is solved by the stepping or asymmetric nature of the aerial gap, but the problem of the high-torque ripple remains unsolved. Generally, two-phase structures best suit high-speed applications and the structures for which the drive and coil cost is an issue. The three-phase structure is one of the most widely used structures, and the four-phase structure is used to reduce torque ripple [1, 2].

Depending on their applications, SRMs are produced within a wide range of structures. **Figure 2** presents a classification of these motors based on their movement patterns, flux path, and type of excitation, each being examined in the following sections.

Figure 1. (a) A three-phase and (b) a four-phase SRM configuration.

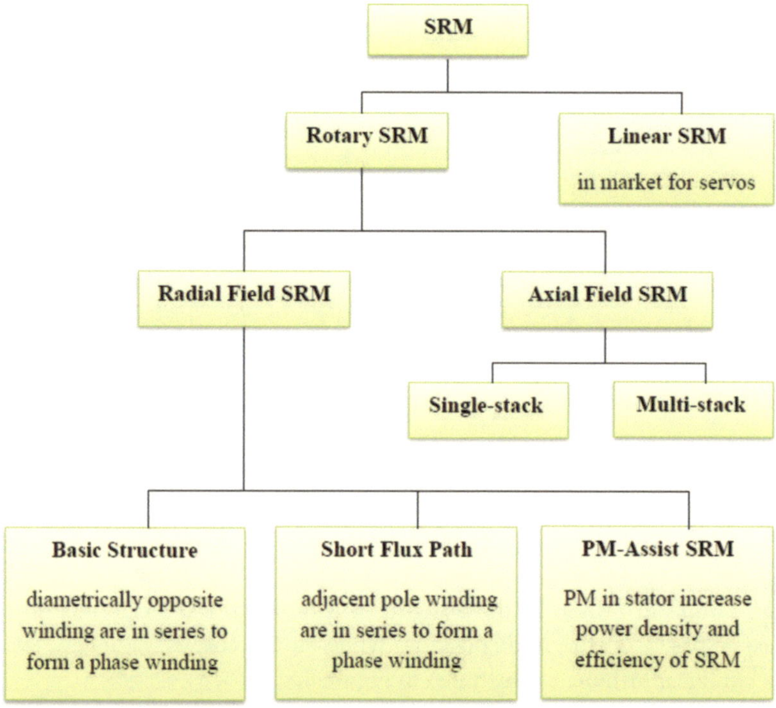

Figure 2. SRM classification based on movement pattern, flux path, and type of excitation.

2. Structural and operational concept of switched reluctance motor

SRMs have a laminated rotor and stator with $N_s = 2 \times m \times q$ poles in the stator and N_r poles in the rotor (m stands for the number of phases and q=1, 2, 3, …). Each phase has a centralized coil on the stator poles. The 6:4 three-phase and 8:6 four-phase structures are among the most common SRM structures (with the first number representing the number of the stator poles and the second number showing the number of the rotor poles), as shown in **Figure 3(a)** and **(b)**.

These two configurations have a constant (q =1), showing that two centralized coils are placed on a pair of poles in each stator phase. Of course, q can be equal to 2 or 3 too, as in 8:12 or 12:18 three-phase configurations which are used in both high-speed, high-torque motors and high-speed generator systems. In addition, to avoid forming zero-torque areas, the same angles ($\beta s = \beta r$) are preferably chosen for the stator and rotor poles [4].

Because of the symmetry of the SRM magnetic circuit, the phase flux linkage is zero even under saturation conditions. Therefore, if a motor phase is short-circuited, the motor is still able to operate with $m - 1$ phases. In this case, due to the lack of mutual induction, no voltage or current is generated in the short-circuited phase. As a result, SRMs are more resistant to faults than other AC motors that operate based on the phase interaction. Besides, self-inductance plays a key role

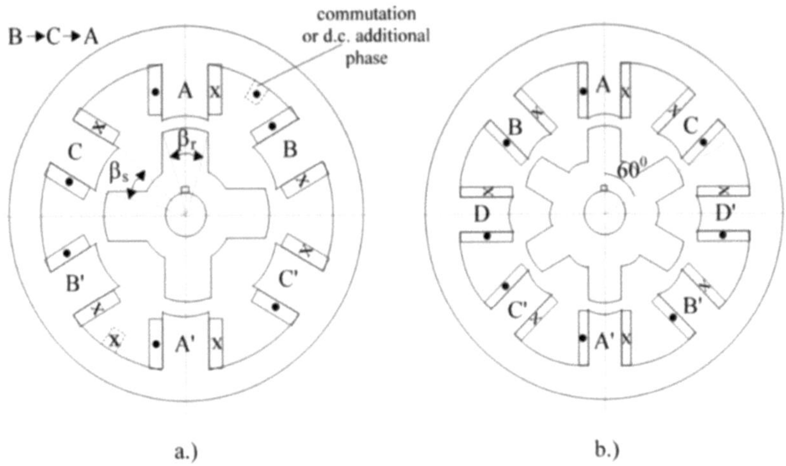

Figure 3. 6:4 three-phase and 8:6 four-phase configurations.

in producing torque in SRMs. In the absence of saturation, self-inductance for each phase changes linearly based on the rotor position, while as the core is saturated, self-inductance changes in a nonlinear fashion, as illustrated in **Figure 4**.

If flux λ is calculated in different rotor positions and is plotted in terms of the current, a class of $\lambda(\theta_r,i)$ curves will be obtained as shown in **Figure 5**. The saturation effect is clearly evident in this figure. Saturation can also be observed even in well-designed motors [4].

If the $W_{mc}(\theta_r)$ co-energy is known, the moment torque of $T_e(i)$ phase can be calculated through Eq. (1):

$$T_e(i) = \left(\frac{\partial W_{mc}(\theta_r)}{\partial \theta_r}\right)_{i=\text{cons.}} ; \; W_{mc} = \int_i^0 \lambda(\theta_r, \, i) di \tag{1}$$

To calculate Eq. (1), the class of $\lambda(\theta_r,i)$ curves needs to be calculated via Eq. (2) as follows:

$$T_e = \sum_{i=1}^{m} T_e(i) \tag{2}$$

The moment torque can be measured through Eq. (3) in cases when there is no saturation:

$$T_e = \sum_{i=1}^{m} \frac{1}{2} i_i^2 \frac{\partial \lambda_i(\theta_r)}{\partial \theta_r} \tag{3}$$

Ideally, when the rotor poles are placed between the two poles of the stator, the phase is excited in the same direction to create motoring function. This is shown in **Figure 6** where the voltage pulse is only applied for conduction angle $\theta_\omega = \theta_c + \theta$.

http://dx.doi.org/10.5772/intechopen.69149

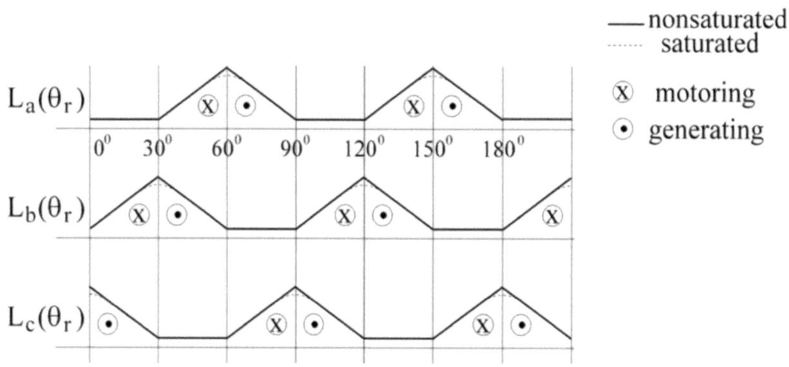

Figure 4. Phase inductance and operation modes in a 4:6 SRM.

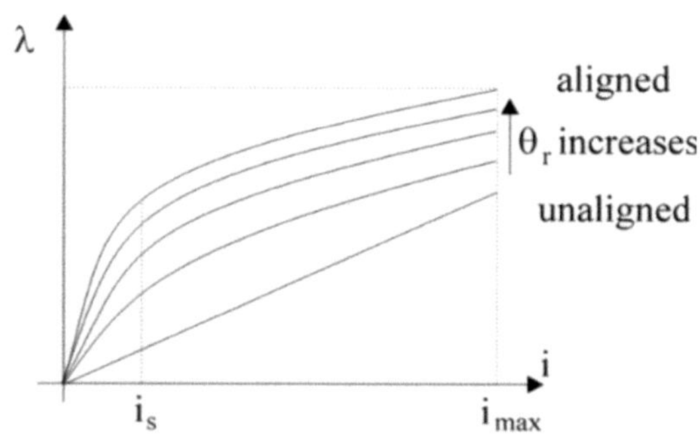

Figure 5. $\lambda(\theta_r,i)$ curves.

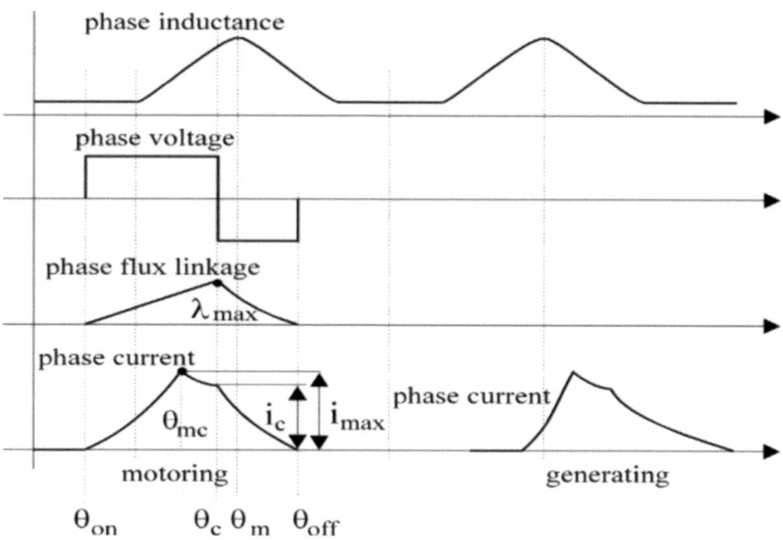

Figure 6. Phase inductance, phase voltage, phase flux linkage and phase current.

Excluding the ohmic voltage drop and taking the speed ω_r as constant, the maximum phase flux linkage (λ_{max}) is calculated as follows:

$$\lambda_{\max} = \int_0^t V_d dt = V_d \frac{\theta_\omega}{\omega_r} \tag{4}$$

The maximum θ_ω for $\theta = 0$ (the zero-pre-phase angle) is determined based on the design using Eq. (5):

$$\theta_{\omega \max} = \theta_m = \frac{\pi}{N_r} \tag{5}$$

The base speed corresponds with $\theta_{\omega \max}$ conditions, single pulse voltage applied with amplitude V_d, and the maximum phase flux linkage. Therefore, it can be suggested that the base speed is dependent on the motor design and the saturation level. As the speed outpaces the base speed, the motor magnetic circuit is saturated.

For speeds higher than the base speed ($\omega_r > \omega_b$) θ_ω should decrease slightly, and subsequently, the maximum phase flux linkage ($\lambda_{\max}$) should be reduced to a certain amount, a phenomenon known as flux attenuation. In addition, in speeds above the nominal speed, in order to achieve the maximum phase flux linkage ($\lambda_{\max}$) at a smaller θ_c angle and the maximum phase flux linkage ($\lambda_{\max}$) at a smaller θ_c angle, and ultimately to generate more torque, the phase firing angle (θ_{on}) should be leading compared to normal conditions. Therefore, the envelope of the torque-speed curve increases accordingly. On the other hand, the phase deactivating process starts at $\theta_c \leq \theta_m$ and ends in the generating zone at θ_{off}. A decrease in $\theta_{\text{off}} - \theta_m$ angle will reduce the share of negative torque in the deactivating process. In practice, if at $\theta_r = \theta_m$ the current value is less than 25–30%, the effect of negative torque will be insignificant [4].

When a phase is cut at an angle θ_c, the other phase turns on and thus the total rate of the torque caused by the interruption of the current in the previous phase is reduced through generating a positive torque.

It is now clear that the magnetic energy of each phase at the conduction time first increases and then decreases. This phenomenon is repeated in each mechanical round for $m \times Nr$ times. At each stage, a part of the magnetic energy is wasted by electronic power converters, and the rest is returned to the DC link and the capacitor of the converter filter. Below the base speed ω_b, current is controlled and restricted by pulse with modulation (PWM) converter as shown in **Figure 7**.

It should be reminded that the conduction time lasts near the angle θ_m, at which the inductance phase angle is maximum. As it was mentioned earlier, the firing phase angle at high speeds is θ_{on} and the turn off angle is θ_c leading.

2.1. The average torque and the energy conversion rate

Magnetic energy conversion levels in both cases of the single voltage pulse (**Figure 6**) and the PWM current and voltage (**Figure 7**) are shown on $\lambda(\theta_r, i)$ curves in **Figure 8(a)** and **(b)**, respectively [4].

http://dx.doi.org/10.5772/intechopen.69149

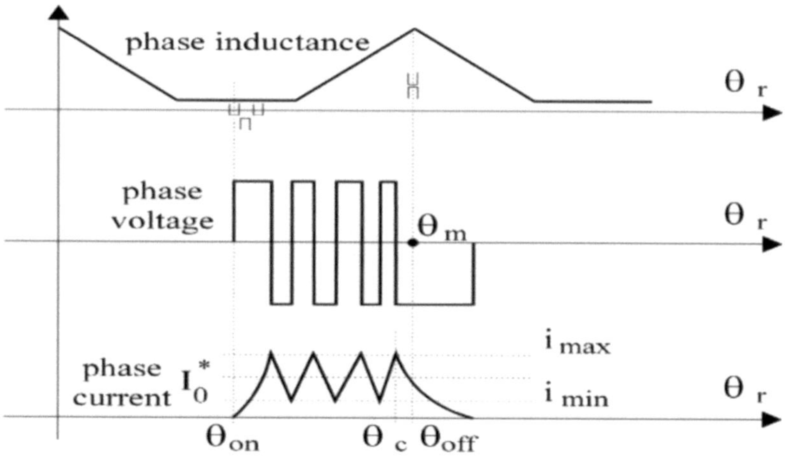

Figure 7. Current and voltage control in a phase below the base speed using PWM converter.

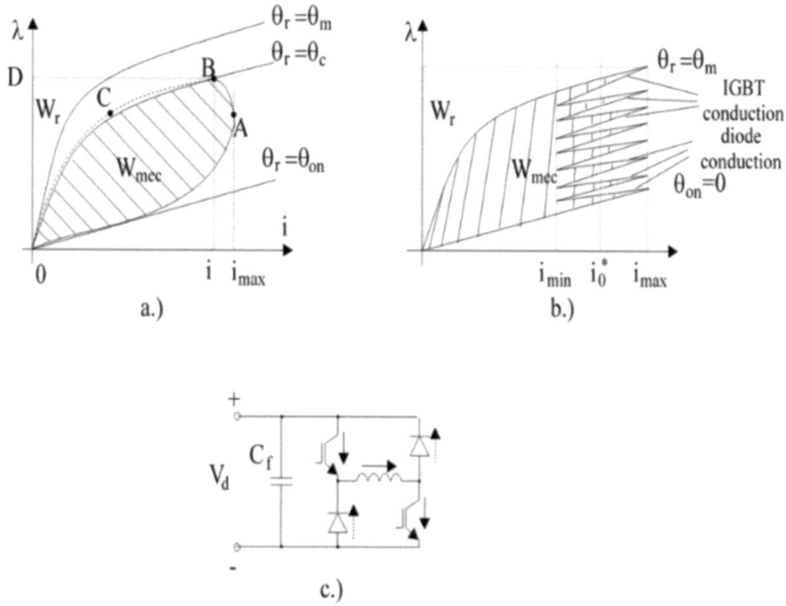

Figure 8. Energy exchange in each phase: (a) at high speeds (single pulse voltage); (b) at low speeds (PWM current and voltage); (c) the converter single-phase feeding configuration to generate a unidirectional current in each phase.

In addition, the converter configuration for feeding a motor phase is shown in **Figure 8(c)**. The average torque (T_{ave}) is proportional to the shaded areas (W_{mec}) in **Figure 9(a)** and **(b)**. Thus, the m-phase SRM average torque and the number of N_r rotor poles at a constant speed can be calculated through Eq. (6):

$$T_{\text{ave}} = \frac{W_{\text{mec}} m N_r}{\theta_c - \theta_{\text{on}}} \tag{6}$$

It should be noted only one phase of SRM conducts at each moment.

2.2. SRM generator function

SRM generator function is slightly more complicated. General principles and strategies for SRM generating mode are described in Ref. [5]. The motor and converters used in this method are shown in **Figure 9**.

To evaluate the generating mode, it is assumed that the generator rotor is rotated anticlockwise by an external torque. In this situation, consider a time when poles 1,1′ are facing poles A_1, A_3 and poles 3,3′ are facing poles A_2, A_4 and the switches S_1, S_2 are closed. In this case, the inductance is at the maximum level and a magnetic flux is generated by the battery in the core. When this happens, the external mechanical torque rotates the rotor and pulls the poles apart, and thus, the inductance is increased. According to minimum reluctance theory, the motor tends to maintain the minimum reluctance, and thus, a torque is generated in the opposite direction of the mechanical torque; thus, the mechanical energy is stored as the magnetic energy in the stator winding. When the rotor reaches a certain angle, switches S_1, S_2 are opened and the current flows through freewheeling diodes into the battery and is stored there [5–7].

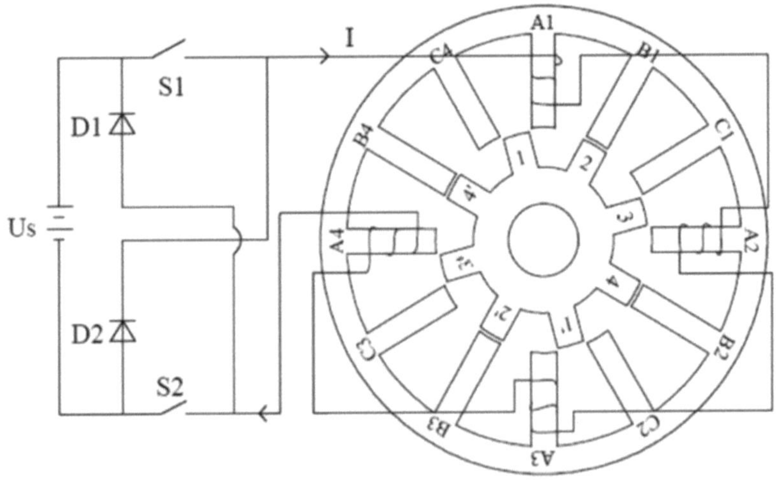

Figure 9. A 12:8 pole three-phase SRM generator configuration.

3. SRM equivalent circuit

Due to the extreme effects of high saturation on $\lambda(\theta_r, i)$ class curves, the mathematical model of the motor is highly nonlinear. However, as the interoperability among phases is insignificant, the cumulative effects of phases torque can be used to calculate the motor torque. An SRM has two saliences in its structure, so we have to use the motor equations in the stator phase frame [4]. The voltage equation is expressed as follows:

$$V_{a,b,c,d} = r_s i_{a,b,c,d} + \frac{d\lambda_{a,b,c,d}(\theta_r, i_{a,b,c,d})}{dt} \tag{7}$$

where $\lambda_{a,b,c,d}(\theta_r, i_{a,b,c,d})$ curves are obtained using curves for a phase periodically with the π/N_s alternation. These curves can be obtained through calculation or experiments. For this purpose, finite element method or analytical techniques can be used. In addition, considering the effects of magnetic saturation and air gap flux distribution is required in all of these techniques. The motion equation is stated as follows:

$$J\frac{d\omega_r}{dt} = T_e - T_{\text{load}}; \quad \frac{d\theta_r}{dt} = \omega_r \tag{8}$$

where

$$T_e = \sum_{a,b,c,d} T_{e_{a,b,c,d}}; \quad \frac{\partial}{\partial\theta_r}\int_0^{i_{a,b,c,d}} \lambda_{a,b,c,d}(\theta_r, i_{a,b,c,d})di_{a,b,c,d} \tag{9}$$

If the subscript i use to refer to the active phase, then Eq. (7) can be rewritten as follows:

$$V_i = r_s i_i + \frac{\partial\lambda_i}{\partial i_i}\frac{di_i}{dt} + \frac{\partial\lambda_i}{\partial\theta_r}\frac{d\theta_r}{dt} \tag{10}$$

The $\frac{\partial\lambda_i}{\partial i}$ term shows the transient inductance (L_t). Thus, we have:

$$L_t(\theta_r, i_i) = \frac{\partial\lambda_i(\theta_r, i_i)}{\partial i_i} \tag{11}$$

The last term in Eq. (10) shows the motoring induction voltage (E):

$$E_i = \frac{\partial\lambda_i}{\partial\theta_r}\omega_r \tag{12}$$

Accordingly, Eq. (10) is rewritten as follows:

$$V_i = r_s i_i + L_t(\theta_r, i_i)\frac{di_i}{dt} + E_i(\omega_r, \theta_r, i_i) \tag{13}$$

Based on this equation, an equivalent circuit with time-dependent parameters can be introduced for SRMs as shown in **Figure 10**.

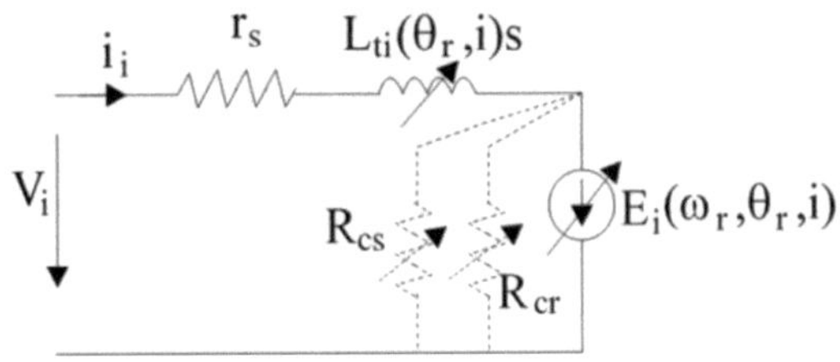

Figure 10. SRM equivalent circuit based on the core loss.

Assuming that the core loss is only due to the main flux component, the core loss can be modeled by variable resistors parallel with the E_i. This motor does not operate based on a rotating field and the core loss occurs in both the rotor and the stator [4].

Taking into account, the energy loss especially at high speeds is required not only to calculate the efficiency but also to evaluate and calculate the transient current response. The operating time of SRMs is generally very high in saturation conditions. In addition, $E(\omega_r, \theta_r, I_i)$ in Eq. (12) is a pseudo emf that contains the terms related to the stored energy. So in this case, the torque should be calculated only through the co-energy equation.

4. SRM categories

As shown in **Figure 3**, SRMs can be categorized into different groups based on their movement patterns and flux paths. This section presents a categorization of SRMs.

4.1. Linear-switched reluctance motors

Linear-switched reluctance motors (LSRMs) are similar to conventional SRMs in their structure except that their rotor and stator are cut open taking a linear form (**Figure 11**). One of the applications of these motors is in electric trains and subways.

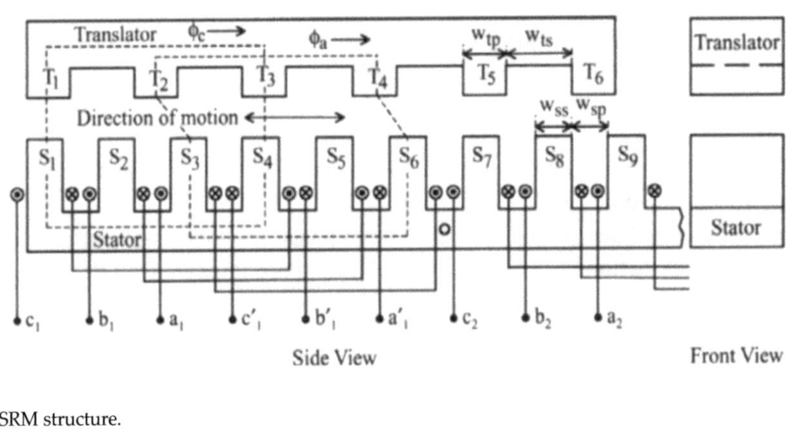

Figure 11. LSRM structure.

4.2. Axial-flux switched reluctance motors

In these motors, the flux path is aligned with the motor axis, and they are used for cases where the motor length is of high importance and the motors with a small length and high torque are preferred for applications such as air conditioning fans and electric vehicles. **Figure 12** displays the configuration of an axial-flux switched reluctance motor.

4.3. Radial-flux switched reluctance motors

Radial-flux switched reluctance motors are the most common structure among SRM structures, and they are divided into two categories:

- Conventional switched reluctance motors: in this structure, the facing poles are connected in series to form a phase.
- Short-path switched reluctance motors: in this structure, the adjacent poles are connected in series to form a phase.

The conventional SRMs are introduced in the first section of this chapter. Other configurations are discussed in the following sections.

4.3.1. Short-flux path SRM

In the short-flux path SRM, the returned flux does not pass through the entire stator yoke, and thus, flux path is shortened, reducing the core loss. **Figure 13** shows a short-flux path SRM. The most common problems with this configuration are the high mutual inductance and the imposition of asymmetric magnetic fractions.

Another type of motors with a short-flux path is the common pole E-core SRM as shown in **Figure 14**. As it can be seen in this figure, this motor has three poles on the stator in which the middle pole lacks winding. In fact, this structure has two phases on the stator, and the middle pole is the shared pole between the two phases and its task is to create a path for the passage of flux. In addition, the shared pole does not play a role in reluctance changing due to its width [8, 9].

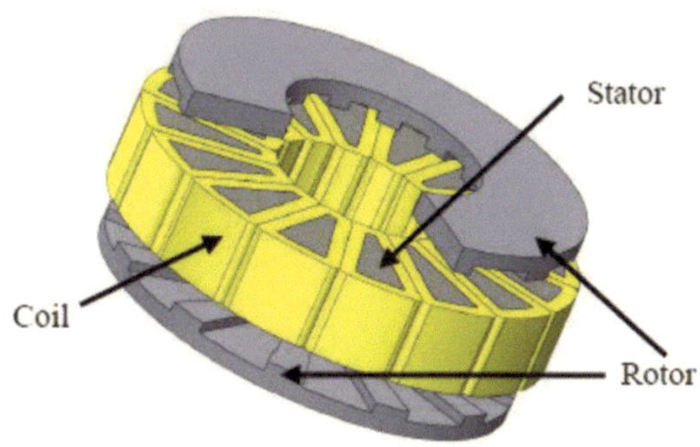

Figure 12. Axial-flux SRM structure.

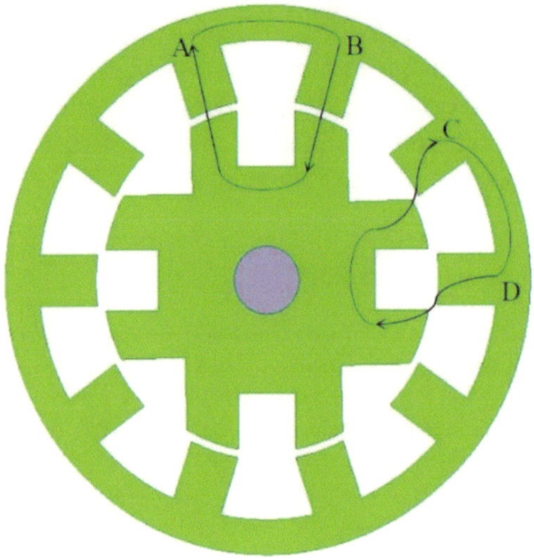

Figure 13. A short-flux path MRS configuration.

Figure 14. A common pole E-core SRM configuration.

Figure 15 displays the functioning and flux path in a common pole E-core SRM. As it is shown, this structure has a short-flux path. To make this type of structure enable to generate torque, at least an E-core must be used in the stator so that it can rotate the rotor. However, two E-cores

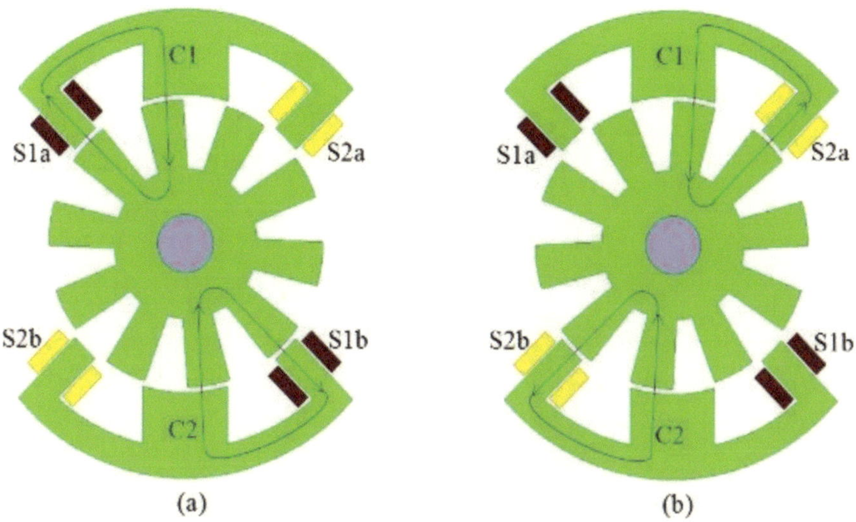

Figure 15. Flux path in a common pole E-core SRM when energizing (a) phase 1; (b) phase 2.

are usually used in the stator because when a single E-core is used it applies an asymmetric axial force to the rotor axis and the motor does not function properly. Therefore, a two E-core stator is employed, as shown in **Figure 15**. Increasing the number of E-shaped parts in the stator increases the torque and power produced by the motor [10–12].

However, a stator with separate parts is not commonly used in common pole E-core SRMs as putting together these parts and preventing their vibration and displacement are a difficult task. Besides, any structural changes may lead to asymmetrical operation. Therefore, the stator of these motors is usually constructed based on integrated 4-E-core motors. **Figure 16** shows a 4-E-core SRM:

4.4. PM-assisted SRM

Because of their own advantages, SRMs have attracted the attention of many industries in recent years. However, these motors suffer from some drawbacks and common problems such as low power and torque density, the complexity of control methods, acoustic noises, and losses related to current excitation. For this reason, some efforts have been made in recent years to use permanent magnetic materials in SRMs. This has led to the emergence of a new configuration called doubly salient permanent magnet motor (DSPMM), which shares the same configuration and functionality of conventional switched reluctance motors and at the same time contains permanent magnet materials in the stator. In many studies, DSPMMs have been considered as a subset of permanent magnet motors. However, as they share the same configuration in the rotor and stator and functionality with SRMs, they have been studied under the category of permanent magnet (PM)-assisted SRMs. This section will examine some of the PM-assisted SRMs, their operating principles, and some of their advantages and disadvantages listed in the literature [2, 13, 14].

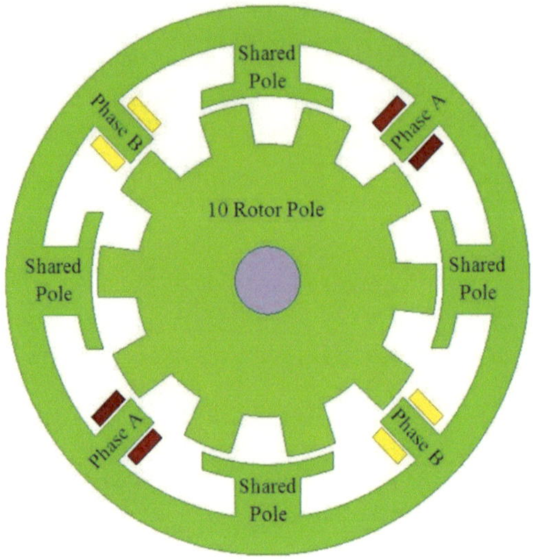

Figure 16. A common pole E-core SRM configuration.

Some of the most important configurations that have been introduced in the literature are as follows:

- Unidirectional flux PM-assisted motors
- Flux-switching motors
- Flux reversal motors
- Hybrid-excited motors

4.4.1. Unidirectional flux PM-assisted motors

This is a common configuration among doubly salient permanent magnet motors (DSPMMs). Even though the rotor and stator are salient poles, the magnetic torque overcomes the reluctance torque, and thus, the generated cogging torque is not considerable. Since with the rotation of the rotor, the linkage flux in each of the coils changes only in one direction, the biggest problem with these motors is the low-torque density caused by the unidirectional flux linkage in each of the coils. **Figure 17** shows two versions of this type of motor [15].

The only advantage of the configuration shown in part (b) compared to the configuration in part (a) is that the former contains more magnetic materials and thus its flux density will be greater. The configuration shown in part (b) is called the yoke linear magnet, and the configuration in part (a) is called the yoke curved magnet [15].

4.4.2. Flux-switching motors

In this configuration, each stator tooth is made of a magnet and two adjacent layers of magnetic core. As shown in **Figure 18**, the stator structure consists of separate U-shaped parts which are

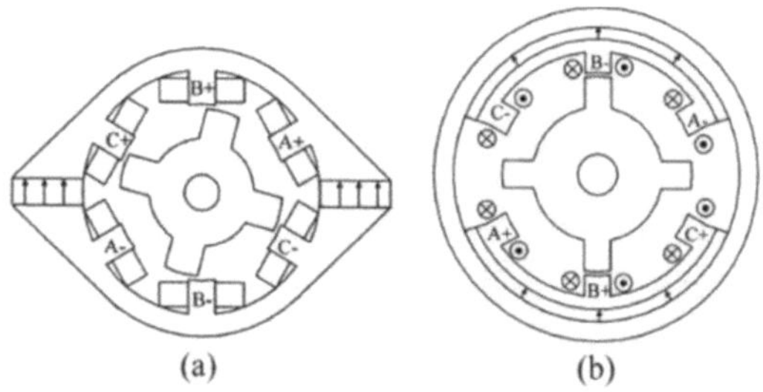

Figure 17. Two unidirectional flux PM-assisted motors.

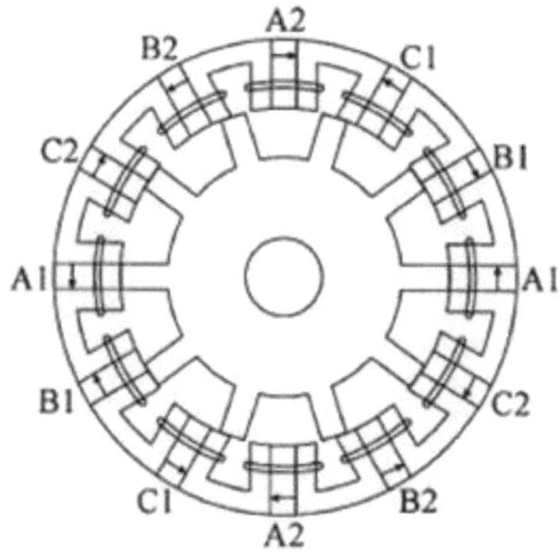

Figure 18. A flux-switching motor.

sandwiched by two adjacent magnets. **Figure 18** shows a constituent cell of this type of motors. A complete motor is formed by putting together a given number of these cells [15, 16].

As shown in **Figure 19(a)**, the flux generated by the magnet is aligned with the flux produced by the coils, and thus, they reinforce each other and, consequently, a great torque is generated. As the rotor rotates to the right, the rotor teeth are moved and the motor is in position (b). In this position, with switching the direction of current in the coil, the fluxes generated by the magnet and the coil are aligned again and they reinforce each other. Therefore, as the direction of the current is switched alternately in the coils, a stable torque will be produced for rotation of the rotor. **Figure 20** displays several other configurations for flux-switching motors [15–18].

4.4.3. *Flux reversal motors*

These motors have a structure that is similar to SRMs with the difference that there is a magnet with different polarity on each pole. Besides, the flux within each coil can be bi-directional in these motors. Since the linkage flux is steadily reversed as the rotor rotates, these motors are called flux reversal permanent magnet motors. In this structure, each stator tooth has a dipole

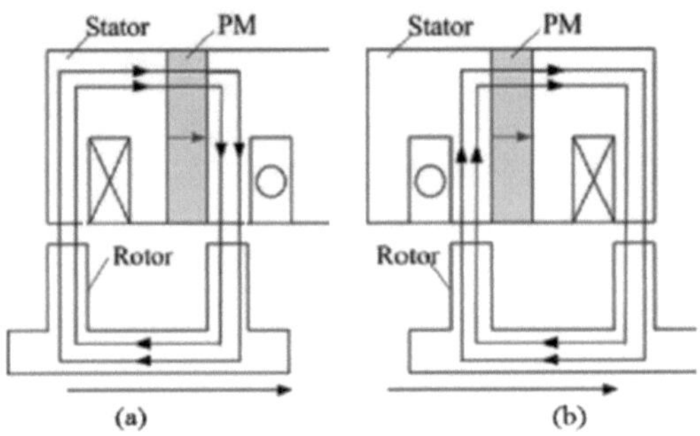

Figure 19. Operational principle of a flux-switching motor.

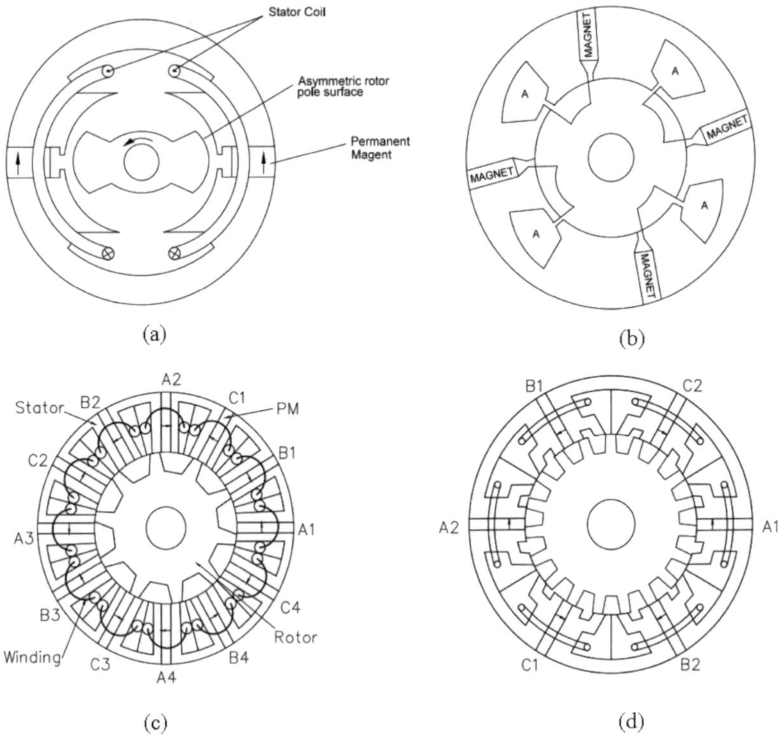

Figure 20. Additional configurations for flux-switching motors. (a) Basic concept (b) Practical one-phase concept (c) Three phase concept (d) Multi-tooth three phase concept.

magnet which is placed on the tooth surface. These motors generate a greater torque because of bi-directional linkage flux of each coil. However, it should be noted that as permanent magnets are placed on the surface of tooth, they are more vulnerable to mechanical damages and may lose their magnetic properties. In addition, there are a significant amount of eddy current losses in the permanent magnet materials. **Figure 21** shows an example of this structure [13]. To explain the principle of operation, a simple single-phase structure of such a motor is illustrated in **Figure 22**.

Figure 21. A flux reversal motor.

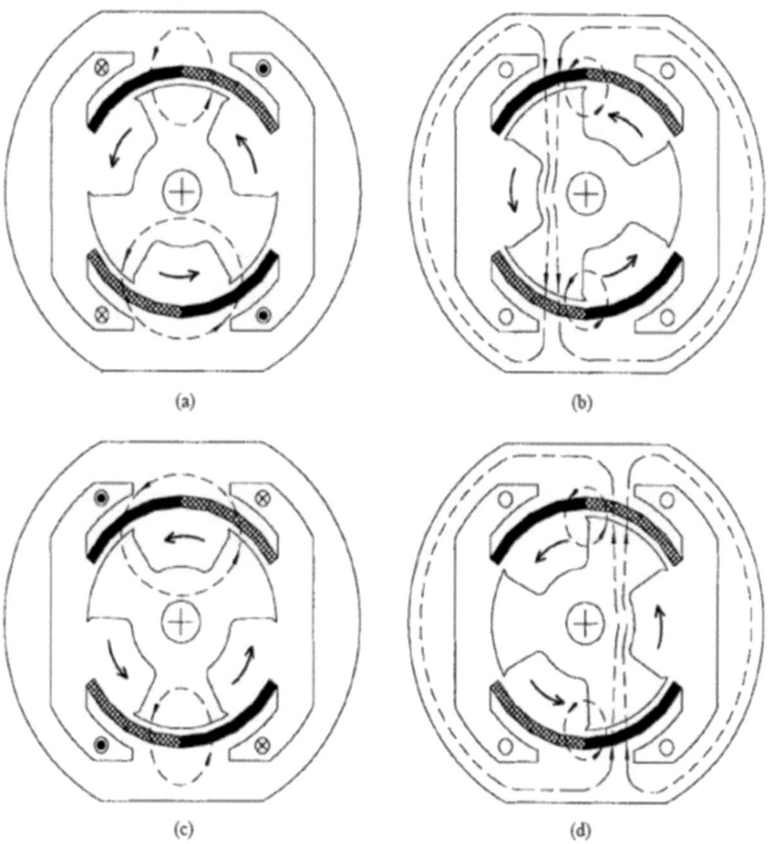

Figure 22. The functioning of flux reversal motors.

Position (a) in this figure is a state of equilibrium in which the rotor remains fixed until the stator windings are excited in the direction of current shown in position (a). The excitation current strengthens the flux of one magnet while attenuates the flux of another magnet which places the rotor in position (b). At this moment, the stator current is disconnected and the rotor moves to the next equilibrium state in position (c). As the current is connected in an opposite direction to position (a), the rotor position is placed in position (d) through the generated flux. The nonstop repetition of these steps will produce torque in the motor. It should be noted that the linkage flux in positions (a) and (c) is at the lowest levels, while it is at the highest level in both positions (b) and position (d) in the positive and negative direction, respectively, as shown in **Figure 22** [13, 15, 19].

4.4.4. *Hybrid-excited DSPM motors*

In this type of motors, the permanent magnet excitation is combined with electrical excitation in the coils and creates an electric field. Hybrid-excited field provides attractive features to these motors and makes them a perfect choice for use in systems such as wind turbines and electric vehicles. Some of these features include:

- The possibility of controlling the air gap flux by changing the polarity and size of the DC current in the excitation coil.
- By strengthening the field, the motor will be able to produce an extremely high torque when necessary.
- By attenuating the field, the constant power zone of the motor is extended and creates a wide speed range.
- By the appropriate adjustment of the air gap flux density, it will be possible to produce a constant output voltage for the generating mode of the motor while the speed of the generator may undergo many changes.
- By the appropriate adjustment of the air gap flux density, it will be possible to create efficiency optimization control (EOC) which in turn makes it possible to optimize the efficiency of the motor during operation. This feature is very prominent because the efficiency of the motor is maintained at an acceptable level at all possible speeds.

Figure 23 shows the structure of a hybrid-excited motor. The structure of the rotor and stator poles in this motor is exactly similar to that of a conventional SRM. However, there are some permanent magnet materials and two coils with DC excitation current in the stator. By changing the DC current in the coils, it is possible to control the excitation flux. This makes it possible to control the motor through simple techniques. This structure is called stator doubly fed doubly salient configuration. Because of the existence of DC excitation coils on the stator, these motors produce a high flux leakage, which is regarded as a serious drawback for these motors [20–22].

4.5. Common pole PM-assist SRM

Ref. [23] introduces a common pole PM-assist SRM, which has a permanent magnet in the excitation pole. **Figure 24** shows the flux path in these motors when energizing phases. As it can be seen in this figure, the flux path is designed in a way that, during energizing each phase,

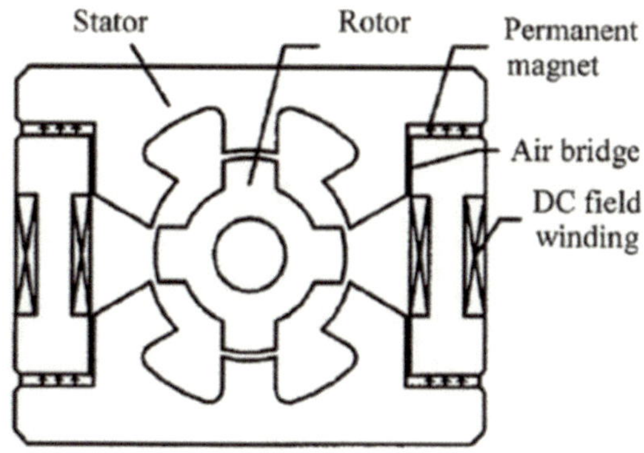

Figure 23. A hybrid-excited DSPM motor.

the fluxes generated by the coil and magnet are aligned with and reinforce each other. One advantage of this structure is that because of the presence of common poles between phases, the flux path does not allow the reverse flux to pass through the permanent magnet, and therefore, there is no risk of losing the magnetic flux of magnets in this structure. In this structure, two magnet blades are placed in excitation poles as shown in **Figure 24**.

One of the effective parameters in increasing the torque produced by the motor is the distance between the two magnet blades. As the distance between the blades increases, the generated torque will also increase. Therefore, the wider the poles, the greater the generated torque will be. Another parameter affecting the torque is the width of the blades as studied in Ref. [23].

Another similar structure is introduced in Ref. [24], with the only difference the blades of the permanent magnet are in the common poles, as shown in **Figure 25**. Given the greater width of the common poles in this structure, the use of magnet blades in these poles is more practical and provides further improvement in the motor performance. These two structures have been studied from different perspectives in Ref. [25]. Generally, it can be concluded that these two structures have a superior performance over the structure that lacks magnet.

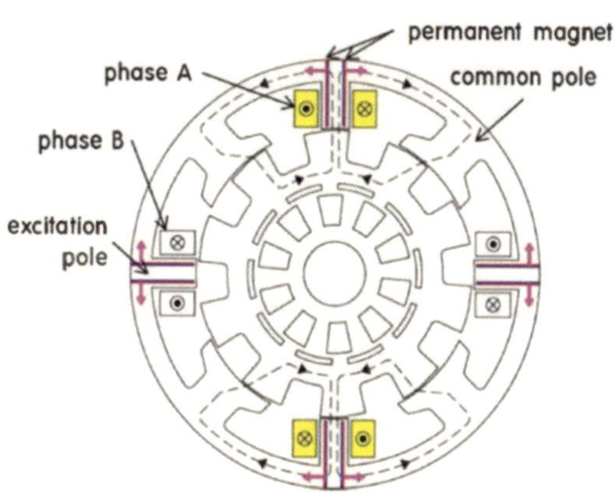

Figure 24. The flux path in a PM-assisted SRM.

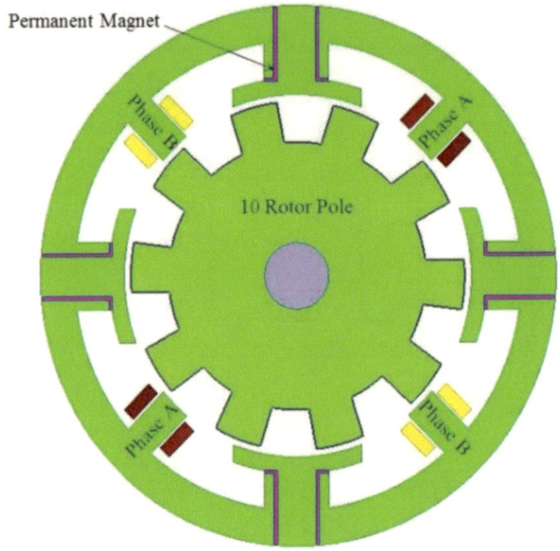

Figure 25. A PM-assisted SRM with a permanent magnet in common poles.

Ref. [8] introduces another configuration of PM-assisted SRMs in which the magnet is placed on the outer surface of the common poles, illustrated in **Figure 26**. The biggest advantage of the structure shown in **Figure 26** is the integrity of the stator compared to the previous structure, which improves the motor strength and performance. But this structure also has a big problem that is related to the location of magnets. If the thickness of the magnet is increased to enhance the flux density, the torque will not increase. This is because as the air gap in the common poles increases, the motor reluctance also increases which decreases the reluctance torque. Therefore, it is possible to use small amounts of permanent magnet materials in this structure.

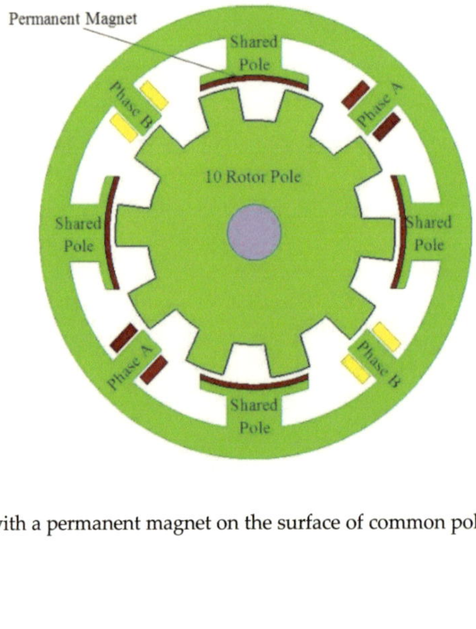

Figure 26. A PM-assisted SRM with a permanent magnet on the surface of common poles.

4.6. Single-phase SRM

Because of their simple structure and low production and maintenance cost, single-phase switched reluctance motors are very appropriate for high-speed applications. When the rotor and stator poles are aligned, the excitation of the stator and rotor poles is interrupted and the rotor continues to move because of the kinetic energy stored in it. When the rotor and stator poles are not aligned, the excitation of the stator windings is resumed and it applies an electromagnetic torque to the rotor. The main problem with this structure is that if the rotor at starting point is in a position where the poles are aligned, the starting torque is not generated and the motor is not able to move. This problem makes these motors inefficient, and they are not used in practice. **Figure 27** shows a very simple single-phase SRM in which the problem of the lack of torque in alignment position is almost solved by placing a permanent magnet on the stator in order to prevent the alignment of the poles in the static mode [4]. These motors have a high-torque ripple and acoustic noise. Thus, they are suitable for use in equipment and tools that are not very sensitive to torque ripple such as home appliances.

Single-phase SRMs have an equal number of poles in the rotor and the stator (2:2 or 4:4). The possibility of a pause in the self-starting position has led to their widespread use in home applications and vehicle parts with a torque less than 1 Nm.

4.7. Double stator-switched reluctance motor (DSSRM)

A qualitative investigation of tangential and normal force densities in electromechanical energy conversion process and the energy conversion process within SRM were presented in [26, 27]. A review of the literature indicates that the majority of the electromagnetic forces that are generated within a conventional SRM are in the radial/normal direction (perpendicular to the direction of motion) and do not contribute to the motion. In fact, a significant part of these forces will initiate undesirable vibrations that have been identified as a major drawback for SRM drives.

It is desirable to generate a larger percentage of the electromagnetic forces that are effectively acting in the direction of motion. Based on these guidelines, double stator-switched reluctance

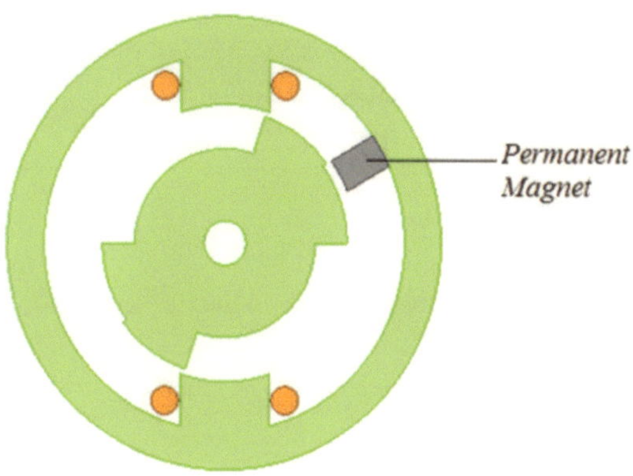

Figure 27. A single-phase SRM with a permanent magnet to prevent the alignment of poles in static mode.

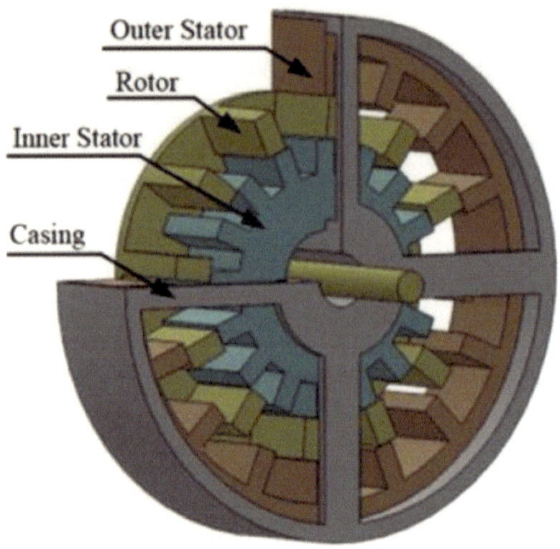

Figure 28. Primary 3-D model of designed DSSCR-SRM.

motor (DSSRM) was proposed in Refs. [28, 29]. **Figure 28** shows the design of a DSSRM proposed in Ref. [29].

5. Conclusion

This chapter presented a comprehensive technology status review highlighting structural and operational concept of SRM, its equivalent circuit model, advantages and drawbacks of each topology as well as recent trends in incorporating PMs in the motor design to boost the overall performance of the motor. The chapter also covers a new class of SRM with double stator geometry. Unlike conventional design in which majority of the electromagnetic force is applied in radial/normal direction (hence not contributing to the motion), double stator design offers a much more efficient configuration in terms of generation of motional forces and exhibits superior performance indexes as compared to conventional design and as such is viewed as a serious contender for high-grade industrial applications.

Author details

Mohammad Mahdi Bouiabady[1], Aliakbar Damaki Aliabad[1] and Ebrahim Amiri[2*]

*Address all correspondence to: eamiri@uno.edu

1 Yazd University, Yazd, Iran

2 University of New Orleans, New Orleans, LA, USA

References

[1] Krishnan R. Switched Reluctance Motor Drives: Modeling, Simulation, Analysis, Design, and Applications. CRC Press, Taylor & Francis Group, Boca Raton, Florida, 2001

[2] Miller TJE. Switched Reluctance Motors and Their Control. Oxford, U.K.: Magna Physics Publishing/Oxford Science; 1993

[3] Rajashekara K. Present status and future trends in electric vehicle propulsion technologies. IEEE Journal of Emerging and Selected Topics in Power Electronics. 2013;**1**(1):3–10

[4] Nasar SA, Boldea I. Electric Drives. Electric Power Engineering Series. 2nd ed. Boca Raton: CRC Press; 2005

[5] GuojunY, et al. Research on the control strategy of switched reluctance generator system. In: Proceedings of the IEEE International in Electronics and Application Conference and Exposition (PEAC); 2014

[6] Bao YJ, Cheng KWE. Experimental examination on a new switched reluctance wind power generator system for electric vehicles. IET Power Electronics. 2012;**5**:1262–1269

[7] Narla S, Sozer Y. Switched reluctance generator controls for optimal power generation and battery charging. IEEE Transactions on Industry Applications. 2012;**48**:1452–1459

[8] Lobo NS. Doubly-salient permanent magnet flux-reversal-free-stator switched reluctance machines. PhD diss, Virginia Polytechnic, Blacksburg, Virginia, 2011.

[9] Lee C, Krishnan R, Lobo N. Novel two-phase switched reluctance machine using commonpole E-core structure: Concept, analysis, and experimental verification. IEEE Transactions on Industry Applications. 2009;**45**(2):703–711

[10] Krishnan R,Lobo NS. Apparatus and method that prevent flux reversal in the stator back material of a two-phase SRM. U.S. Patent 7 015 615, March 21, 2006

[11] Oh SG, Krishnan R. Two phase SRM with flux reversal free stator: Concept, analysis, design and experimental verification. In: Proceedings of the Conference Record of the IEEE IAS Annual Meeting, Vol. 3; 2006. pp. 1155–1162.

[12] Eskandari H, Mirsalim M. An improved 9/12 two-phase e-core switched reluctance machine. IEEE Transactions on Energy Conversion. 2013;**28**(4):951–958

[13] Deodhar RP, et al. The flux-reversal machine: A new brushless doubly-salient permanent-magnet machine. IEEE Transactions on Industry Applications. 1997;**33**(4):925–934

[14] Liao Y, Liang F, Lipo TA. A novel permanent-magnet motor with doubly-salient structure. IEEE Transactions on Industry Applications. 1995;**31**:1069–1078

[15] Chau K, Chan C, Liu C. Overview of permanent-magnet brushless drives for electric and hybrid electric vehicles. IEEE Transactions on Industrial Electronics. 2008;**55**(6):2246–2257

[16] Cheng M, Chau KT, Chan CC. Design and analysis of a new doubly salient permanent magnet motor. IEEE Transactions on Magnetics. 2001;**37**(4):3012–3020

[17] Hoang E, Ben Ahmed A, Lucidarme J. Switching flux permanent magnet poly-phased synchronous machines. In: European Conference on Power Electronics and Applications; 1997

[18] Zulu A, Flux switching machines using segmental rotors. PhD Thesis. Newcastle University, Newcastle upon Tyne, Tyne and Wear, UK, 2010

[19] Deodhar RP, Andersson S, Boldea I, Miller TJE. The flux-reversal machine: A new brushless doubly-salient permanent magnet machine. IEEE Transactions on Industry Applications.1997;**33**(4):925–934

[20] Chau K, et al. Design and control of a PM brushless hybrid generator for wind power application. IEEE Transactions on Magnetics. 2006;**42**(10)3497–3499

[21] Fan Y, Chau KT, Cheng M. A new three-phase doubly salient permanent magnet machine for wind power generation. IEEE Transactions on Industry Applications. 2006;**42**(1):53–60

[22] Chau KT, Jiang JZ, Wang Y. A novel stator doubly fed doubly salient permanent magnet brushless machine. IEEE Transactions on Magnetics. 2003;**39**(5):3001–3003

[23] Hwang H, Hur J, Lee C. Novel permanent-magnet-assisted switched reluctance motor (I): Concept, design, and analysis. In: Proceedings of the 2013 International Conference on Electrical Machines and Systems (ICEMS); 2013

[24] Jeong J, Her J, Lee C. Novel permanent-magnet-assisted switched reluctance motor (II): Concept, design,and analysis. In: Proceedings of the 2013 International Conference on Electrical Machines and Systems (ICEMS); 2013

[25] bookCho J, Hur J, Lee C. Control scheme of a novel permanent-magnet-assisted switched reluctance machine. In: Proceedings of the 2013 International Conference on Electrical Machines and Systems (ICEMS); 2013

[26] Jiang W, Moallem M, Fahimi B, Pekarek S. Qualitative investigation of force density components in electromechanical energy conversion process. In: Proceedings of the 32nd Annual Conference on IEEE Industrial Electronics (IECON); November; 2006. pp. 1113–1118

[27] Edrington CS, Kaluvagunta DC, Joddar J, Fahimi B. Investigation of electromagnetic force components in SRM under single and multiphase excitation. IEEE Transactions on Industry Applications. 2005;**41**(4):978–988

[28] Abbasian M, Moallem M, Fahimi B. Double-stator switched reluctance machines (DSSRM): Fundamentals and magnetic force analysis. IEEE Transactions on Energy Conversion. 2010;**25**(3):589–597

[29] Asgar M, Afjei E. Radial force reduction in a new flat-type double-stator switched reluctance motor. IEEE Transactions on Energy Conversion. 2016;**31**(1):141–149

Chapter 2

Four-Quadrant Control of Switched Reluctance Machine

Sandeep Narla

Additional information is available at the end of the chapter

http://dx.doi.org/10.5772/intechopen.69228

Abstract

This chapter illustrates modeling techniques and software simulation of the switched reluctance machine (SRM) machine models with controllers for efficient operation. The first model is based on torque and flux data generated through finite element analysis (FEA) and the second model is geometry-based machine model, which are used to develop the operation logic for four-quadrant control of an SRM. The results obtained from these models were used to develop a control strategy to adapt turn-on and turn-off (commutation) angles efficiently. Two digital controllers, namely the phase current controller for regulating current with a hysteresis band and the PI (proportional-integral) speed controller for regulating the speed, are developed to deliver the desired output torque. The controller is based on a negative feedback closed-loop control system.

Keywords: SRM, four-quadrant, motoring, generating, FEA, modeling, simulation

1. Introduction

Earlier chapters may have described the fundamental principles of the SRM such as its stator, rotor configuration, construction, and basic operation. The shape of the inductance profile of a switched reluctance machine (SRM) along the air-gap between stator and rotor poles depends on the resistance along the gap and pole-widths of stator and rotor as shown in **Figure 1**. The control algorithm is developed based on this inductance profile with respect to rotor position. Motoring can be achieved between unaligned and aligned pole positions because of the positive slope of inductance. Similarly, along the negative slope from aligned to unaligned pole positions, generating torque can be produced [1].

During motoring operation, the machine produces positive torque between unaligned and the next aligned position in forward or reverse directions. In generation, the inductance will be

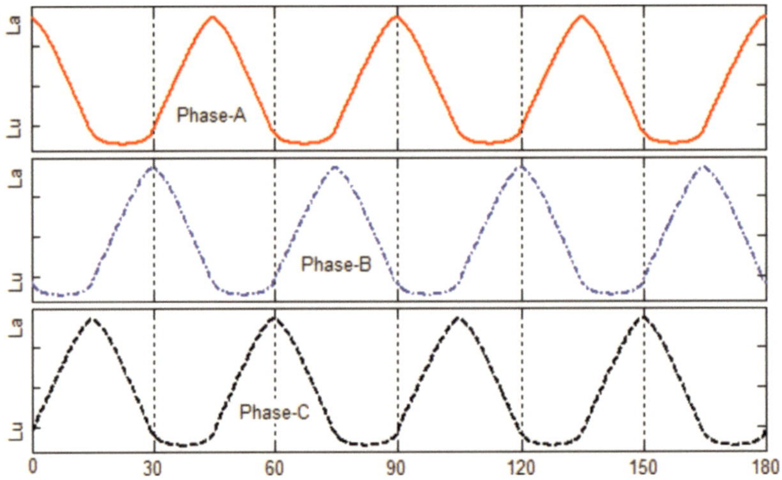

Figure 1. Phase inductance profile of a 12/8 pole SRM with respect to rotor position (in mechanical degrees).

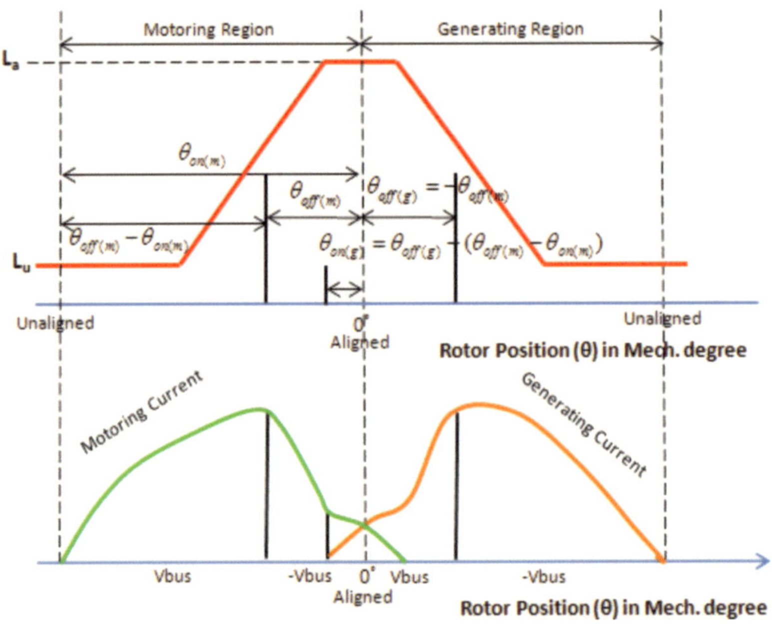

Figure 2. Motoring and generating current waveforms describing the phase inductance principle of SRMs.

decreasing in the direction of rotation resulting in negative torque. The current flow in the phase winding is always unidirectional; however, the current profiles for the motoring and generating modes are exactly the mirror image of each other along the aligned pole position for the same speed and symmetrical turn-on and turn-off angles as shown in **Figure 2.**

2. Classic converter topology

Since the torque developed in an SRM is independent of the direction of current flow, unipolar converters are sufficient to serve as the power converter circuit for the SRM. The most flexible and versatile four-quadrant SRM topology is the classic bridge converter, which has two transistors and two freewheeling diodes per phase as shown in **Figure 3**. The transistor switches are turned on and off in each phase based on the controller output for torque and speed control of the SRM.

There are three modes of converter operation in each phase: magnetization, freewheeling, and demagnetization.

1. Magnetization period for motoring is when both the switches are turned on and the energy is transferred from the source to the motor phase winding as shown in **Figure 4(a)**. For generation, the phase winding of the motor is excited initially to generate higher back-emf than the DC-link voltage ($+V_{dc}$) with the machine driven by a prime mover. The generated energy is delivered to the electrical side achieving generation.

2. Freewheeling, for motoring operation at lower speeds, is accomplished by keeping one of the switches on and switching the other switch as shown in **Figure 4(b)**. With only one switch on, the motor phase gets slowly demagnetized through the respective antiparallel freewheeling diode. For generating operation, one of the switches is turned off, whereas the other switch is turned on and off. With only one switch on, the motor phase gets slowly magnetized through the respective antiparallel freewheeling diode.

3. Demagnetization is achieved by applying a negative DC-link voltage to the phase winding with the two switches turned off as shown in **Figure 4(c)**, helps in fast decay of current flowing through both the diodes.

The chopping operation involves switching of the transistors connected to the phase winding accordingly for current regulation depending on the motoring or generating operation. The possible phase voltages in this Pulse Width Modulation (PWM) operation would be $+V_{dc}$ (for magnetization), 0 V (freewheeling), and $-V_{dc}$ (for demagnetization). The phase current control is set by three levels, which include the reference current value "I_{ref}" and the upper and lower hysteresis band limits. The hysteresis band value is chosen based on the peak current value of the machine. The chopping operation is implemented for torque control below the based speed.

The single-pulse mode of operation is possible when the machine is operated at high speeds. The possible phase voltages in this operation would be $+V_{dc}$ (for magnetization) and $-V_{dc}$ (for demagnetization). The current regulation is not possible in this operation because in motoring there may not be sufficient time to reach the desired current level at high speeds or in generating, when demagnetization occurs, there is no control over the diodes. The main advantage of this converter is its ability for independent control of each phase, which is particularly important when phase overlap is desired. It is more suitable for high-voltage, high-power drives.

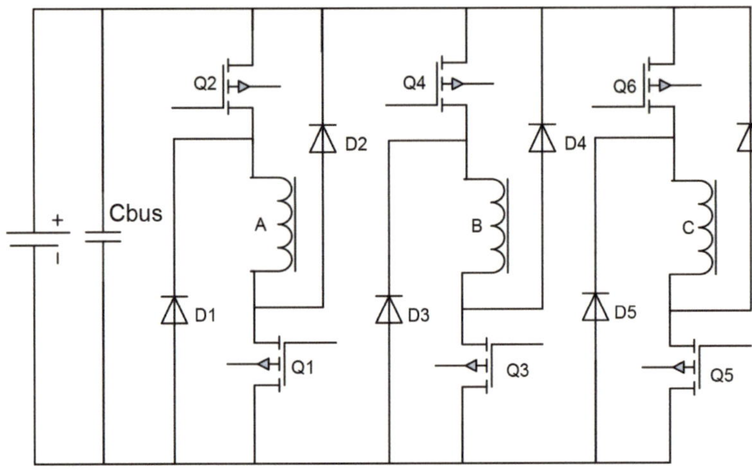

Figure 3. Four-quadrant classic converter for a three-phase SRM.

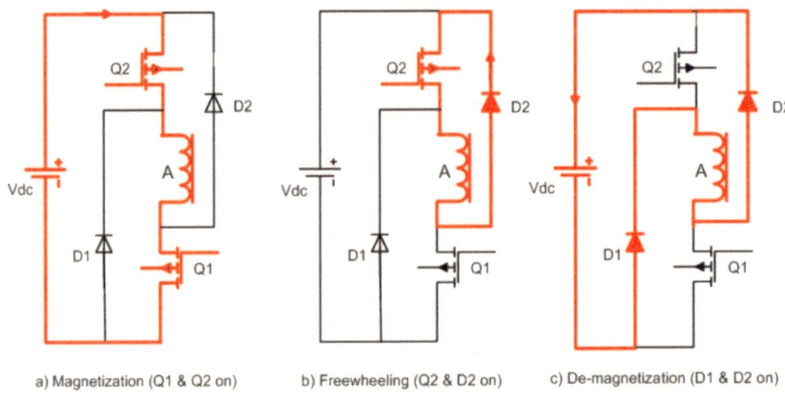

Figure 4. Converter switching operation for motoring of phase A winding.

3. Controller strategy

The effective performance characteristics from an SRM drive system can be obtained by proper positioning of the phase excitation pulses relative to the rotor position. The commutation angles (turn-on angle "θ_{on}," turn-off angle "θ_{off}"), total conduction period, and the magnitude of the phase current "I_{ref}" determine the average torque, torque ripple, and other performance parameters. The complexity of finding the control parameters depends on the chosen control method for a particular application. At low speeds, the current rises almost instantaneously after turn on because of the negligible back-emf and the current must be limited by either controlling the average voltage or by regulating the current level. As the speed increases, the back-emf increases and opposes the applied DC-link voltage. Phase advancing is necessary to establish the phase current at the onset of rotor and stator pole overlap region. Voltage PWM is used to force maximum current into the machine to maintain the desired torque level.

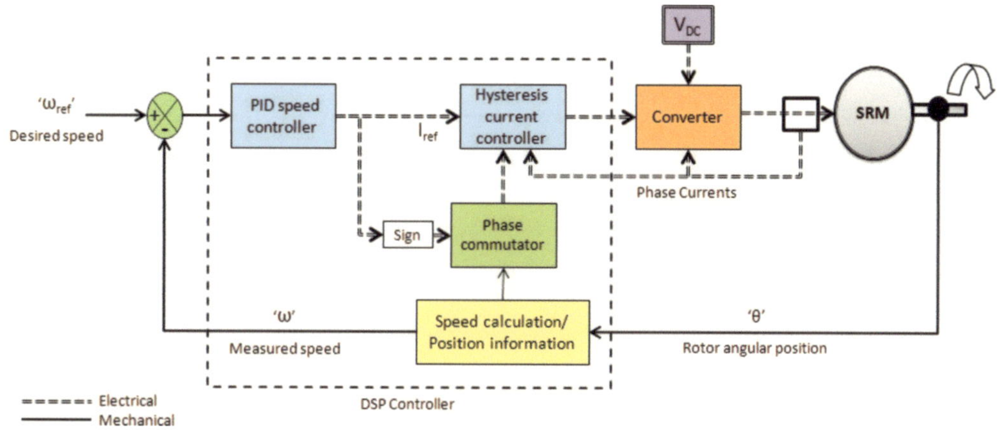

Figure 5. Closed-loop speed and current control block diagram of a SRM.

The block diagram for the general closed-loop speed and current control of the SRM is shown in **Figure 5**. In speed control applications, an outer speed loop is added to the faster inner current loop. The inner loop creates desired currents in the stator windings necessary to achieve the desired speed specified by the outer loop. The inner current loop is implemented by current regulation, which consists of the current controller, the power converter, the SRM stator windings and the current sensing devices for feedback control. For current control, a simple hysteresis controller is illustrated here for simplicity and PI controller for outer speed control.

4. Modeling of SRM in finite element analysis

Flux 2D is a finite element analysis software by Magsoft Corporation [2] that allows the user to determine the various characteristic parameters of a machine based on the mechanical dimensions, material properties, and operating conditions of the machine [10]. The overall cross-section of the machine is divided into many sections, such as shaft, rotor, air-gap, stator, and so on, and individual properties are defined. At the intersections of each section and on the surface, the desired machine parameters of the SRM are calculated by virtually running the machine at a constant speed in small position steps. The finite element analysis gives the flux-linkage and torque characteristics of the SRM with respect to rotor position and current level. These characteristics are used to build the SRM model in the form of look-up table. The developed look-up table model is used in the simulation of the machine model.

An Finite Element Analysis (FEA) package requires detailed input data as shown in **Table 1** for design and the results need skilled interpretation. The SRM operates in a series of strokes with "switched mode" excitations having no steady-state reference where all its state variables are constant. The initial sizing of the geometry, static and magneto-dynamic FEA and system-level simulations should be considered simultaneously for the SRM design to meet the design requirements [3]. Other design objectives can be to optimize the torque density, power density, efficiency and to minimize torque ripple.

Parameters	Value	Units
Shaft radius	0.6	Inches
Radius to rotor yoke	1.2025	Inches
Radius to air gap	1.6325	Inches
Radius to stator yoke	2.233	Inches
Outer radius	2.7435	Inches
Air-gap	0.015	Inches
Number of phases	3	–
Number of stator poles	12	–
Number of rotor poles	8	–
Number of repetitions	2	–
Periodicity	90	Degrees
Rotor pole width	15	Degrees
Stator pole separation	30	Degrees
Magnetic saturation	1.8	Tesla
Magnetic permeability	4pi × e-7	Henry/meter
Relative permeability	3300	–
Fill factor	0.453	–
Stacking factor	0.9	–
Length of stack	1.846	Inches
Phase resistance	2.1	Ohms
No. of turns/pole	175	–
No. of coils in series	4	–
No. of coils in parallel	1	–
Max. current in coil	5.5	Amperes
Power delivered	660	Watts
Operating speed	264	Radians/sec

Table 1. Dimensions, configuration parameters, and ratings of an example 3-ph, 12/8 SRM.

Modeling of SRM in Flux 2D involves eight steps: (a) geometry construction, (b) sectional design, (c) surface meshing, (d) stator and rotor material, (e) stator coil excitation circuit, (f) rotor physics definition, (g) solving for machine parameters, and (h) analysis of results [4].

5. Simulation of the SRM FEA model in Mat-lab

Look-up tables are developed for the machine model from the flux-linkage and torque characteristics obtained from FEA. The flowchart showing the SRM simulation with FEA look-up

http://dx.doi.org/10.5772/intechopen.69228

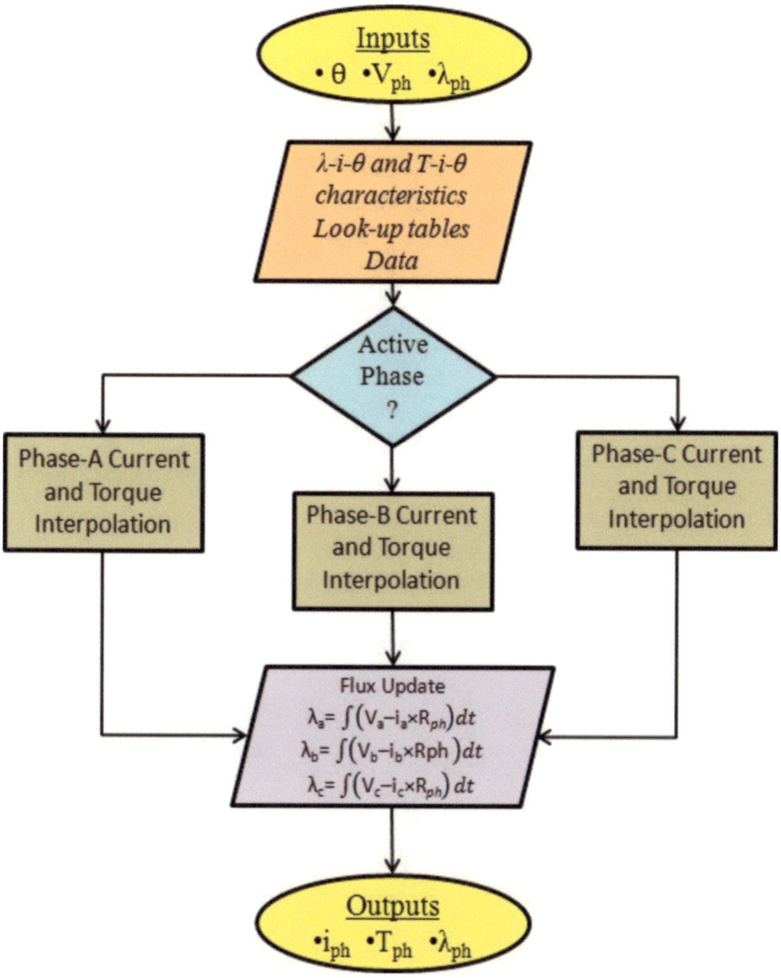

Figure 6. Flowchart showing the SRM simulation with FEA look-up table-based model.

table-based model is shown in **Figure 6**. The control logic for this simulation is similar to the geometry-based machine model described below. The table for flux-linkage characteristics consists of flux values at different current levels and rotor positions as shown in **Figure 7**. The table for torque characteristics consists of phase torque values for different current levels and rotor positions as shown in **Figure 8**. Depending on the mode of operation (motoring or generating), the respective rotor positions are considered appropriately.

5.1. Phase current interpolation

In the flux-linkage table, the columns represent the phase current (in amperes) and the rows represent the rotor position (in degrees). The flux-linkage values at different rotor positions are represented as rows between unaligned and aligned positions. Now depending on the rotor position and the flux-linkage values, the two adjacent current columns are known. Finally, the

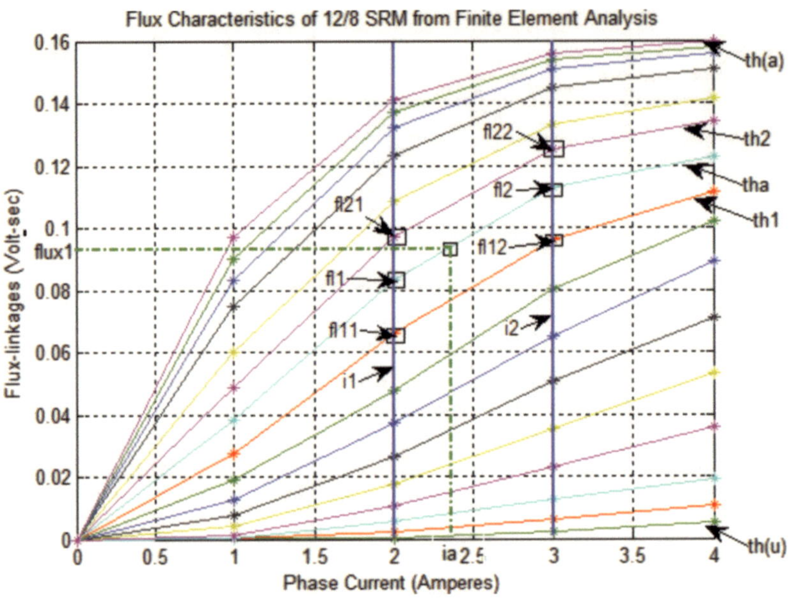

Figure 7. Flux-linkage characteristics obtained from Flux2D finite element analysis.

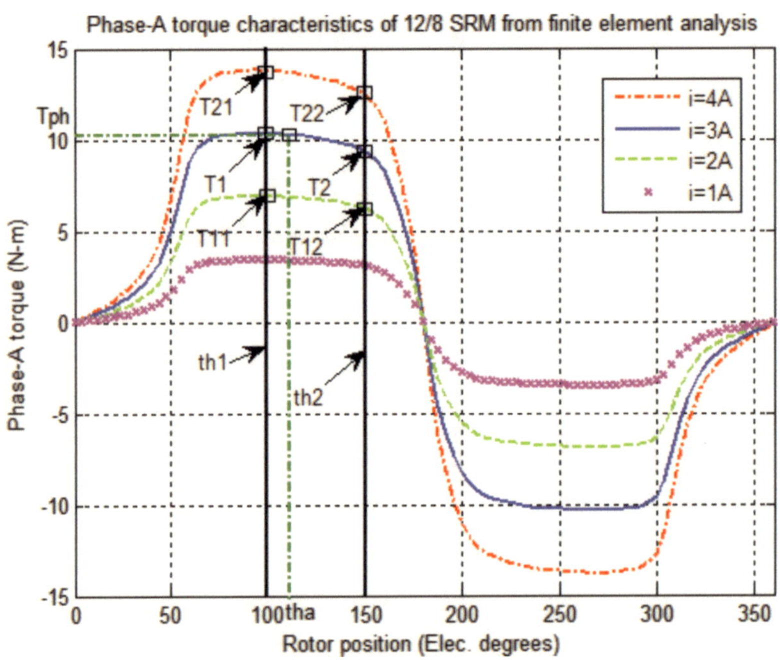

Figure 8. Torque characteristics obtained from Flux2D finite element analysis.

phase current is interpolated from this data as shown in Eqs. (1)–(8). The interpolation algorithm is described below.

A set of four fluxes is obtained first from available data in flux-linkage table. These are

$$fl_{11} = flux(row_1,\ col_1) \tag{1}$$

$$fl_{12} = flux(row_1,\ col_2) \tag{2}$$

$$fl_{21} = flux(row_2,\ col_1) \tag{3}$$

$$fl_{22} = flux(row_2,\ col_2) \tag{4}$$

The interpolated fluxes fl_1, fl_2 are obtained from fl_{11}, fl_{12}, fl_{21}, fl_{22}, respectively, using the equation:

$$fl_1 = (fl_{21} - fl_{11}) \times (\ th_1 - th_1)/(\ th_2 - th_1)\ + fl_{11} \tag{5}$$

$$fl_2 = (fl_{22} - fl_{12}) \times (\ th_1 - th_1)/(\ th_2 - th_1) + fl_{12} \tag{6}$$

Here, th_a is in mechanical degrees.

The final phase current at fixed "th_a" is calculated as:

$$i_{ph} = (i_2 - i_1) \times (phase\_flux - fl_1)/(fl_2 - fl_1) + i_1 \tag{7}$$

After determining the appropriate phase current value, the rate of change of flux-linkage is computed based on the equation below. By integrating the above equation, the instantaneous phase flux-linkages can be obtained.

$$d\lambda_{ph}/dt = V_{ph} - (i_{ph} \times xR_{ph}) \tag{8}$$

5.2. Phase torque interpolation

Similar to phase current, the phase torque is also interpolated using the same procedure [12]. The interpolation equations with reference to the **Figure 8** below are given in Eqs. (9)–(11).

$$T_1 = (T_{21} - T_{11}) \times (th_a - th_1)/(th_2 - th_1) + T_{11} \tag{9}$$

$$T_2 = (T_{22} - T_{12}) \times (th_a - th_1)/(th_2 - th_1) + T_{12} \tag{10}$$

Here, th_a is in mechanical degrees.

The final phase torque at fixed "th_a" is calculated as:

$$T_{ph} = (T_2 - T_1) \times (i_{ph} - i_1)/(i_2 - i_1) + T_1 \tag{11}$$

6. Modeling of SRM in Mat-lab

The SRM is always operated in the magnetically saturated mode to maximize the energy transfer. Knowledge of the magnetic flux linked by a phase is essential to develop a sophisticated SRM controller. The inherent magnetic non-linearity of the SRM must be taken into account by accurate modeling of the machine characteristics [5]. The high degree of non-linearity makes it [9] impossible to model the flux-linkage or phase inductance by an SRM phase accurately. The phase inductance and flux-linkage vary with rotor position due to stator and rotor saliencies and also vary with the instantaneous phase current because of magnetic saturation.

The analytical model developed by A. Radun is derived from the machine geometry and material magnetic property. The model uses an analytical solution for the flux linked and static torque produced by one SRM phase. Separate analytical models for the flux linked by a phase when its stator and rotor poles do overlap [6] and do not overlap [6] are combined to provide a complete model of a given SRM phase. When the poles overlap, saturation must be included, especially in the pole tips, whereas a linear representation can be used for the unsaturated regions. The flux linked by the SRM phase is determined from the sum of the main flux and the fringing flux that is linked to the phase. The analytical model could be used to calculate motor parameters.

The geometry-based machine model allows extensive computer simulation studies during the machine and drives design stage. The general form of a geometry-based analytical expression for flux-linkage used can be shown as [6]:

$$\lambda(i,\theta) = A_m(\theta,\xi) + A_f(\theta,\xi) - B_m(\theta,\xi) \times sqrt\big(C_m(\xi) + D_m(\theta,\xi) + E_m(i^2,\xi)\big) - B_f(\theta,\xi) \times sqrt\big(C_f(\xi) + D_f(\theta,\xi) + E_f(i^2,\xi)\big) \tag{12}$$

Where, A, B, C, D, and E are dependent constants. "ξ" stands for geometry and magnetic

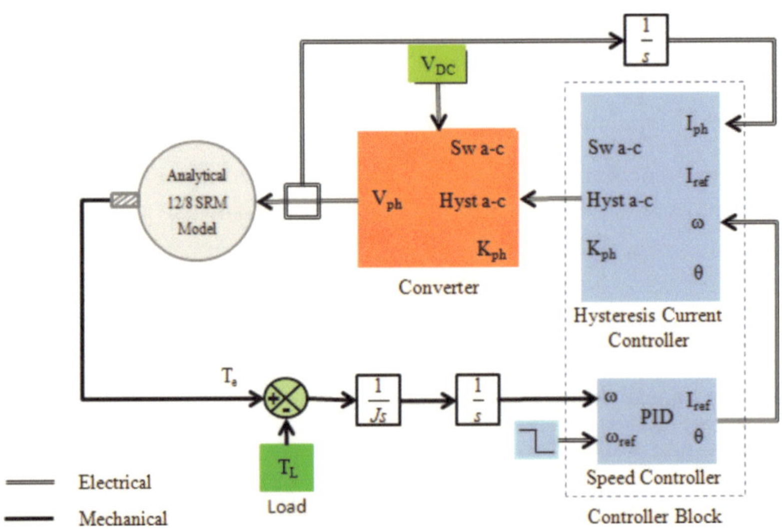

Figure 9. Simulation block diagram of analytical SRM model.

properties and "θ" stands for rotor position. Subscripts "m" and "f" are for main and fringing components of the flux-linkage, respectively. This approach is important for modeling of the magnetic structure of the motor for the purposes of controller design and dynamic performance prediction.

The above model described is used in the mat-lab simulation [7]. The block diagram of the simulation is shown in **Figure 9**. The modules for the simulation are the machine model, the speed and current controllers, the power converter, and the load model. The inputs to the machine model are the phase voltages, and the outputs are the phase currents, rotor position, and electromagnetic torque. In the machine model, the current derivatives *di/dt* are calculated based on the machine parameters, applied phase voltages, and the previous step phase currents and position information. The equations to calculate the phase torque, phase inductance, back-emf and phase currents have been derived from the machine model. The fundamental inputs used for this simulation are the geometric parameters and operating conditions of the machine.

7. Simulation of the SRM analytical model in Mat-lab

The controller generates the turn on, turn off, and phase current command based on a control algorithm and rotor position feedback information [11]. The sequence of phase turn-on and turn-off logic for each phase is simplified by wrapping the angles for once rotor pole period, i.e., -22.5 to 22.5° based on the 12/8 pole SRM configuration shown in **Figure 10**. Since the machine has 12 stator and 8 rotor poles, the pole arc of each rotor pole is 45°; therefore, the position is periodic within -22.5 to 22.5° or 22.5 to -22.5° depending on the direction of rotation (forward or reverse directions, respectively). Similarly, the position wrapping for phase *B* and *C* are shifted by 30 and 60° (mechanical) respectively [8].

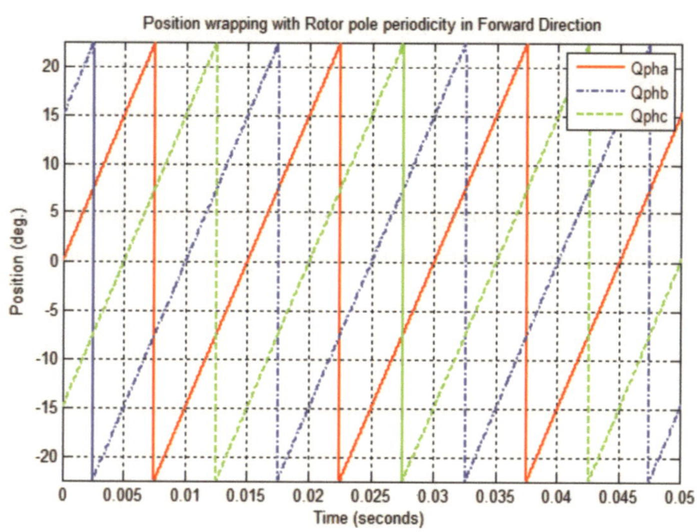

Figure 10. Position wrapping of each phase with rotor pole arc period.

7.1. Motoring operation of the SRM

The best turn-on and turn-off angles are tuned for proper current and speed control based on the inductance profile of the machine [13]. The various operations involved in modeling the closed-loop control of the SRM for motoring are active phase determination, phase switching, hysteresis current chopping, and PI speed control.

a. Active phase determination: The active phase to be turned on for power processing is based on the rotor position. Phase A aligned position is used as the reference (initialized to 0°); therefore, phase C will be starting phase for motoring operation if going in forward direction of motion or phase B if going in reverse direction. When the phase is operating in its increasing inductance region and within the conduction period, it is said to be active. The turn-on and turn-off angles may change, and the period of phase operation may also vary depending on the torque demand and operating speed. The nominal phase conduction period is the stroke angle, which is 15° for a 12/8 SRM.

b. Phase commutation: In order to achieve precise control of phase current at various machine operating speeds, each active phase has to be operated between the optimum turn-on and turn-off angle positions. When the phase operation starts, the winding is excited until the current command is reached and then demagnetized when the phase operation is turned off. Chopping can also be implemented for torque control.

c. Current chopping: Once the phase is excited with both switches turned on $+V_{dc}$ is applied, the phase current will increase as the phase voltage dominates over the back-emf resulting in positive *di/dt* as described by Eq. (13):

$$di/dt = (V_{ph} - back\_emf - i \times R_{ph})/L_{inc}(\theta) \tag{13}$$

When the current reaches the upper band [I_{ref} + *band*/2], one of the switches is turned off, whereas the other switch is kept on for the phase current to decrease by freewheeling within the phase winding because of negative *di/dt*. When the current reaches the lower band [I_{ref} – *band*/2] due to 0 V, both the switches are turned on to increase the phase current. Both the switches are turned off when the phase has to commute upon approaching the turn-off angle so that the phase current decays to zero quickly as shown in **Figure 12**. During motoring operation, the phase voltage applied across the phase winding by switching of transistors is shown in **Figure 11**. This procedure is repeated for next in-coming phases.

The machine rotor speed "ω_m" and position "θ" is computed with the known moment of inertia "J" the total electromagnetic torque "T_e" developed by the machine, and the opposing load torque "T_L" using the equation;

$$J \times d\omega/dt = T_e - T_L \tag{14}$$

d. PI speed controller: A PI controller is used for speed control of the SRM for simplicity. The advantage with this controller is that the speed of the machine can be controlled with fast

http://dx.doi.org/10.5772/intechopen.69228

transient without overshoots and good stead state response. The PI speed control loop is in the outer loop generating the desired phase current command based on the speed as shown in closed-loop block diagram **Figure 5**. Here, the desired current "I_{ref}" is not fixed, but changes based on the error between desired speed "ω_{ref}" and actual machine speed "ω". The PI control parameters are tuned in the simulation **Figure 12**.

7.2. Generating operation of the SRM

Here, the SRM is mechanically coupled and driven by the prime mover at a constant speed. The various operations involved in modeling the braking operation of the SRM are active phase determination, phase commutation, and hysteresis current chopping [14].

a. Active phase determination: Generation is achieved during decreasing inductance region $-dL/d\theta$ from aligned to unaligned positions. The conduction angle in this region is chosen based on the control algorithm for desired braking operation. The same turn-on and turn-off angles are used for other phases with phase angle shifting. The turn-on and turn-off angles may change depending on the speed and generation current command.

b. **Phase commutation:** The phase operation in generation is quite different from motoring operation. When the phase switching is turned on, the winding is initially excited for some period and then generation is achieved with the excitation turned off. When the excitation is off, the phase winding is demagnetized through the diodes in order to capture the generated energy. Chopping mode with hysteresis current control can also be used for regeneration in the lower speed range.

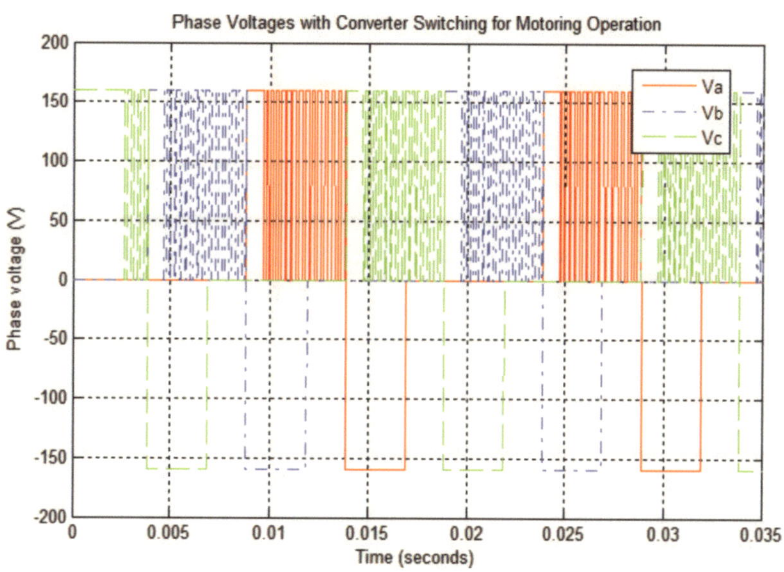

Figure 11. Phase voltage switching during motoring operation at fixed speed.

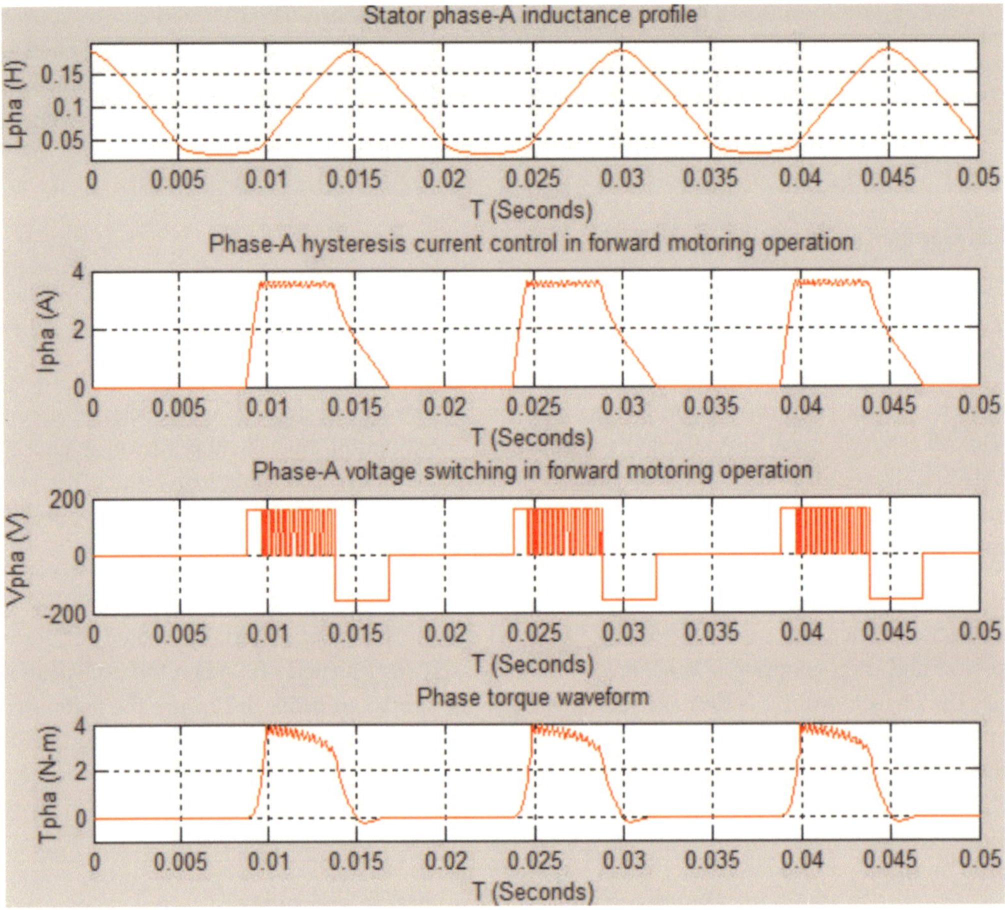

Figure 12. Forward motoring operation of phase A with hysteresis current control.

c. Current chopping: The phase current control is implemented same as in motoring, but the freewheeling of phase currents is achieved differently. The back-emf produced during the generation is negative; therefore, when the voltage across phase winding is 0 V, the rate of current change is positive resulting in magnetization. Similarly, with the phase voltage $-V_{dc}$ the rate of current change is negative resulting in current decay based on Eq. [13].

The phase winding is initially excited for some period by applying $+V_{dc}$ across the phase. When the phase excitation is turned off, the phase current will continue to increase even after the excitation period is finished and the current regulation follows until the phase operation is commuted. During current regulation, when the current reaches the upper band $[I_{ref} + band/2]$, both the switches are turned off ($-V_{dc}$) and when the current reaches the lower band $[I_{ref} - band/2]$, one of the switches is turned on (0 V) to freewheel the current within the phase winding. When the phase has to commute, both the switches are turned off ($-V_{dc}$) to let the phase current decay to zero soon and then set to 0 V, when the current reaches zero as shown in **Figure 13**.

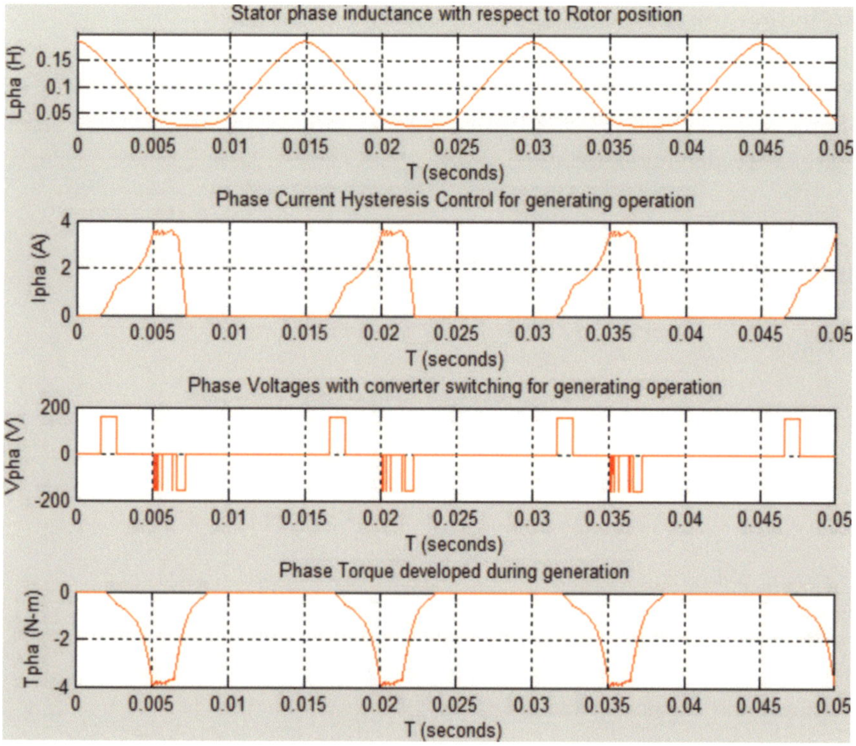

Figure 13. Generating operation of the SRM in forward direction.

d. Single-pulse mode operation: When the machine is operated above the based speed, it is forced to be in the single-pulse mode. In this mode, the phase winding is initially excited with both the switches turned on ($+V_{dc}$) and then demagnetized with both the switches turned off ($-V_{dc}$). Here, the current cannot reach the desired level and decreases to zero when approaching phase commutation point.

8. Conclusion

A Switched Reluctance Machine is special DC machine with non-linear characteristics. This chapter provides modeling approach of two different techniques, one based on the analytical equation and other based on finite element analysis. The techniques provide guidance on how to obtain the optimized commutation angles from the modeling and by utilizing them in the software simulation, the closed-loop control of the SRM in motoring and generating operation are demonstrated. With the optimized commutation angles, precise control of speed, torque, and current in various operational modes in four-quadrants like forward motoring, reverse motoring, forward generation and reverse generation are shown.

Acknowledgements

I wish to express my deep sense of gratitude and indebtedness to my academic advisors, Dr. Yilmaz Sozer and Dr. Iqbal Husain, for their great support in the course of research and opportunity to broaden my horizons.

Author details

Sandeep Narla

Address all correspondence to: narlasandeep@gmail.com

The University of Akron (Alumni), Akron, Ohio, USA

References

[1] Miller TJE. Switched reluctance motors and their control. New York: Magna Physics Publishing and Clarendon Press, Oxford; 1993

[2] Magsoft Corporation. Finite element analysis software reference manual. Troy, New York: Magsoft Corporation; 2010

[3] Radun AV. Design considerations for the switched reluctance motor. IEEE Transactions on Industry Applications. 1995;**31**(5):1079–1087

[4] Dawson GE, Eastham AR, Mizia J. Switched reluctance motor torque characteristics: Finite element analysis and test results. IEEE Transactions on Industrial Applications. 1987;**30**(3):532–537

[5] Husain I, Hossain S. Modeling, simulation and control of switched reluctance motor drives. IEEE Transactions on Industrial Electronics. 2005;**52**(6):1625–1634

[6] Hossain S, Husain I, Klode H, Kequesne B, Omekanda A. Four-quadrant control of a switched reluctance motor for a highly dynamic actuator load. Applied Power Electronics Conference and Exposition, APEC. 7th Annual IEEE, Vol 1. 2002

[7] Hossain S, Husain I. A geometry based simplified analytical model of switched reluctance machines for real-time controller implementation. IEEE Transactions on Power Electronics. 2003;**18**(6):1384–1389

[8] Radun AV. Generating with the switched reluctance motor. In: Proceedings of the IEEE Applied Power Electronics Conference; Dayton, OH, USA; 1994. pp. 41–47

[9] Krishnan R. Switched reluctance motor drive: Modeling, simulation, analysis, design and application. USA: Magna Physics Publishing; 2001

[10] Koibuchi K, Ohno T, Sawa K. A basic study for optimal design of switched reluctance motor by finite element method. IEEE Transactions on Magnetics. 1997;**33**(2):2077–2080

[11] Radun AV. Analytical calculation of the switched reluctance motor's unaligned inductance. IEEE Transactions on Magnetics. 1999;**35**(6):4473–4481

[12] Moallem M, Ong CM. Predicting the torque of a switched reluctance machine from its finite element field solution. IEEE Transactions on Energy Conversion. 1990;**5**(4):733–739

[13] Radun AV. Analytically computing the flux linked by a switched reluctance motor phase when the stator and rotor poles overlap. IEEE Transactions on Magnetics. 2000;**36**(4):1996–2003

[14] Torrey DA. Switched reluctance generators and their control. IEEE Transactions on Industrial Electronics. 2002;**49**:3–14

Direct Instantaneous Torque Controlled Switched Reluctance Motor Drive for Fan Type Load and Constant Torque Load

Srinivas Pratapgiri

Additional information is available at the end of the chapter

http://dx.doi.org/10.5772/intechopen.69280

Abstract

Switched reluctance motor (SRM) drives can be a good competitor to conventional induction and permanent magnet motors in variable speed applications because of advantages, such as simple construction, no rotor windings, high torque to inertia ratio, adaptability to hostile conditions, etc. Due to its high nonlinearity, the torque ripple is high in switched reluctance motor. The sophisticated direct instantaneous torque control (DITC) can maintain the torque developed by the motor within hysteresis band by suitably selecting the switching states of the converter. Hence, DITC controller minimizes the torque ripples and also provides fast response to the torque changes. The performance of DITC controlled SRM drive is analyzed through simulations during acceleration and also in steady state for two types of load torques namely fan type and constant torque. The variation of the switching frequency of the converter is analyzed by changing the torque hysteresis band. It has been observed that as the hysteresis band decreases, the switching frequency increases. So, the hysteresis band cannot be increased beyond a certain limit so as to ensure that the switching frequency of the device cannot increase beyond its operating limit.

Keywords: direct instantaneous torque control, switched reluctance motor, torque ripples

1. Introduction

Switched reluctance motors (SRMs) are replacing the conventional motors in specific applications during last decade due to developments in power electronics and microelectronics technology. Due to its simple mechanical structure, manufacturing process becomes easy and the cost of the motor becomes low. The rotor is made up of stacks of iron and does not carry any windings or magnets. Thus, it is mechanically robust and naturally suited for high-speed operation. The rotor has the low moment of inertia and high torque/inertia ratio. The SRM achieves high-torque levels at low-peak currents by using small air gaps. The major sources of heat are on the stator, so simple cooling methods can be adopted, as the stator is easier to access than the rotor. The rotor losses are much smaller than the stator as compared to DC, induction and PM machines. As the motor does not produce cogging or crawling torque, skewing is not required to decrease them unlike the induction and PM machines [1–4].

The SRM is a highly reliable motor, as it can function even under faulty conditions with reduced performance. In addition, the motor windings are both physically and electromagnetically isolated from one another, reducing the possibility of phase-to-phase faults. As the windings are electrically separate from each other and as they have negligible mutual coupling, electrical fault in one phase does not affect other phases. Such a feature is unique to the switched reluctance motor. The classic SRM drive as a system with converter involves two switches and a winding in series. Thus, even in case of both switches being turned on simultaneously, no shoot-through faults would occur, unlike the case of AC drives, which lead to shorting of the DC bus [3]. This assures high reliability for this converter compared to other converters. Unidirectional current required by the motor drive makes power electronics drive circuitry simple and reliable.

Main drawback of the motor is that the torque ripples are high as compared to DC and AC machines. Developments in semiconductor technology and microelectronics have resulted in development of sophisticated control strategies to minimize the torque ripples. This has brought the SRM back into the variable-speed drive market. To compete for applications requiring no position sensor, advanced sensor-less techniques have been suggested by the researchers [4]. Applications of the SRM drives are categorized as low, medium, high-power, and high-speed drives [3, 5]. The low-power (less than 3 HP) applications are plotter drive, washing machines, hand-held tools, sliding doors and household appliances (mixer machines, vacuum cleaners with a power range of 0.5–2 kW). The applications in medium power range (less than 300 kW) are industrial general purpose drives, electric vehicles, trains, mining industry; speed applications are textile processing industry, centrifuge for medical applications and aerospace applications [5].

2. Switched reluctance motor

The switched reluctance motor is a doubly salient and singly excited motor. This means that it has salient poles on both stator and rotor as shown in **Figure 1**. The stator poles carry the concentric coils. The diametrically opposite coils on the stator poles are connected to form the phase winding. The rotor has no windings and is built up from a stack of salient pole laminations [6]. The motor operates on the well-known principle of minimum reluctance. When a particular phase is excited, the nearest rotor pole tries to align along the corresponding stator pole axis, so as to minimize the reluctance of the magnetic path. Hence, by sequential

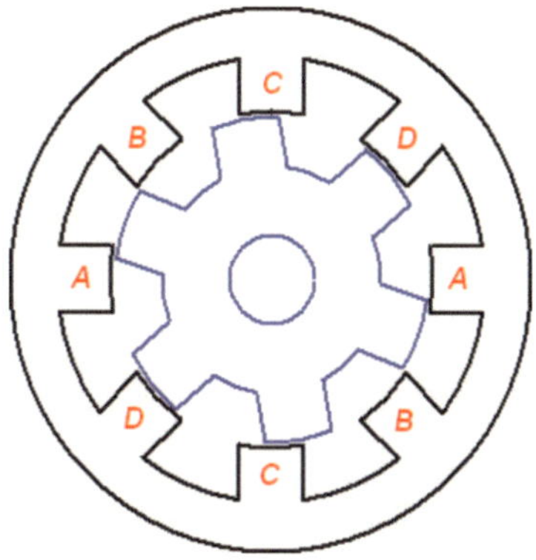

Figure 1. A typical four phase 8/6 SRM.

excitation, the motor can rotate in either direction. An exciting sequence of A-B-C-D for the motor shown in **Figure 1** will result in counter clockwise rotation, while an exciting sequence of C-B-A-D will result in clockwise rotation of the motor [6]. During the rotation, there are three relative positions between the stator and the rotor. When any pair of the rotor poles is exactly aligned with stator poles of a particular phase, that position is called as *aligned position*. Similarly, if the inter-polar axis of the rotor is aligned with the stator poles of a particular phase, that position is called as *unaligned position*. The position between the *aligned* and *unaligned position* is called the *misaligned position* [7].

The mathematical equations of DITC [8, 9] as applied to SRM are discussed here. The instantaneous voltage across the motor winding is given by:

$$v = Ri + \frac{d\psi(\theta, i)}{dt} \tag{1}$$

where $\psi(\theta, i)$ is the phase flux-linkage, which is a function of rotor position θ and current i.

Thus, the equation for the power flow can be written as:

$$vi = Ri^2 + i\frac{\partial\psi(\theta, i)}{\partial i}\frac{di}{dt} + i\frac{\partial\psi(\theta, i)}{\partial\theta}\frac{d\theta}{dt} \tag{2}$$

$$dW_{\mathrm{m}} = i\frac{\partial\psi(\theta, i)}{\partial\theta}d\theta - \frac{\partial W_{\mathrm{f}}}{\partial\theta}d\theta \tag{3}$$

where, dW_{m} and dW_{f} are the differential mechanical energy and field energy, respectively.

The instantaneous torque is defined by:

$$T = \frac{dW_{\mathrm{m}}}{d\theta} \tag{4}$$

The expression for the instantaneous torque production of an SRM phase can be written as:

$$T = i\frac{\partial\psi(\theta,i)}{\partial\theta} - \frac{\partial W_f}{\partial\theta} \tag{5}$$

This is a rarely used variant of conventional torque equation. Due to saturation in the SRM, the influence of the second term in Eq. (5) is negligible. Therefore, by using this approximation, the following equation for torque production may be obtained as:

$$T \approx i\frac{\partial\psi(\theta,i)}{\partial\theta} \tag{6}$$

3. Principle of direct instantaneous torque control

Direct instantaneous torque control strategy as shown in **Figure 2** can be explained as follows [10–13]. The reference torque is compared with the actual torque and given to a hysteresis torque controller, which outputs in torque increase or decrease signal. The output of the torque hysteresis is given to switching control unit. Switching control unit does the following operations: it detects the outgoing and incoming phases based on the position sensor signal, turn-on and turn-off angles of the converter. Next states of both incoming and outgoing phases are determined on the information obtained from instantaneous torque, command torque, and present states of incoming and outgoing phases so as to control the torque within its set band. The asymmetrical converter used to excite the phases of the motor has three possible switching *States*, i.e., 1, 0, and −1 [11, 14].

In single-phase conduction, any one phase is excited, and the total torque is developed by that phase only. If the torque is less than the reference torque, the controller selects *State* as 1 and if it is more than the reference torque *State* is made to −1. *State 0* is not used in this scheme. During phase commutation, two phases conduct simultaneously, and the torque of the two consecutive phases is controlled to control the total torque. The torque is maintained within

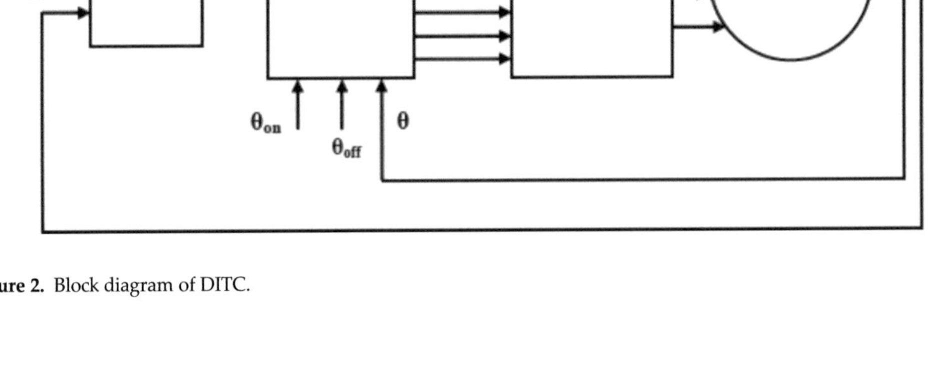

Figure 2. Block diagram of DITC.

http://dx.doi.org/10.5772/intechopen.69280

hysteresis band, by changing the *States* of the outgoing and incoming phases between 1 and −1, depending on the value of instantaneous torque [11, 14]. The flow chart of the DITC is shown in **Figure 3**.

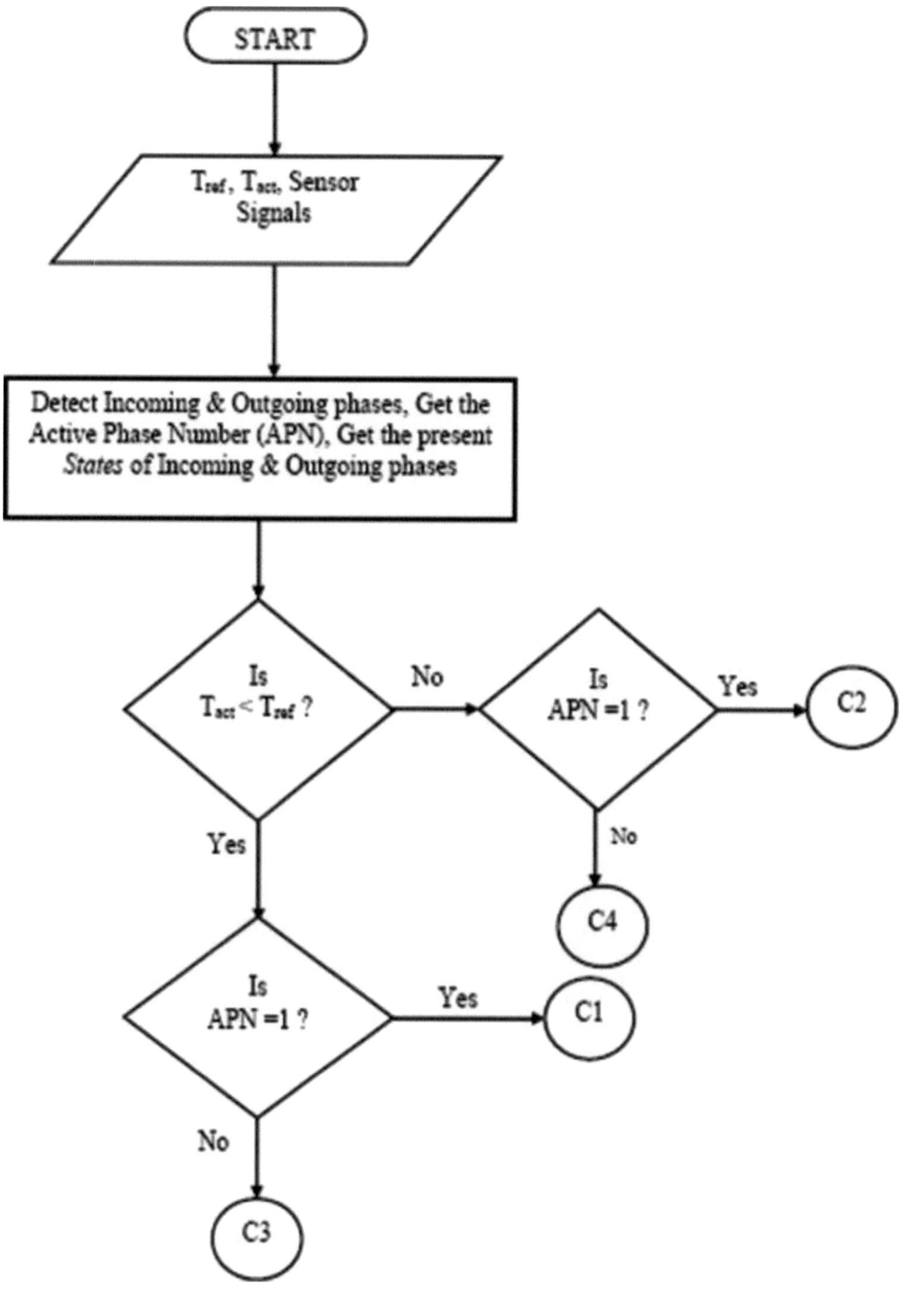

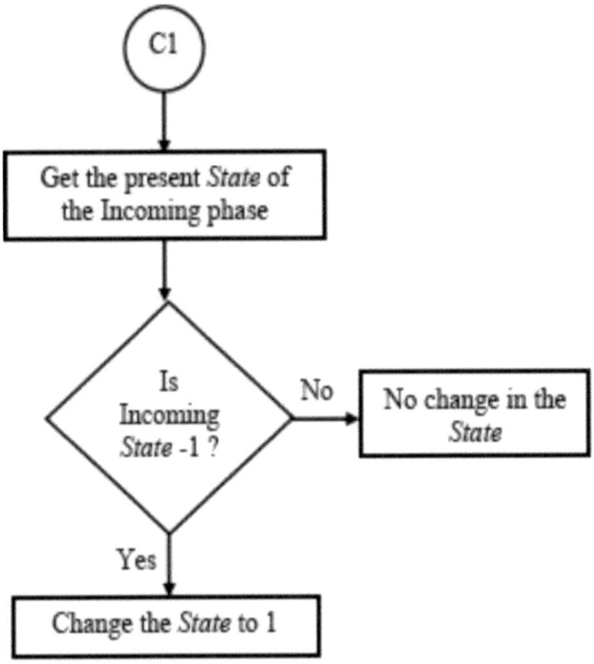
C1
Get the present State of the Incoming phase
Is Incoming State -1 ?
No
No change in the State
Yes
Change the State to 1

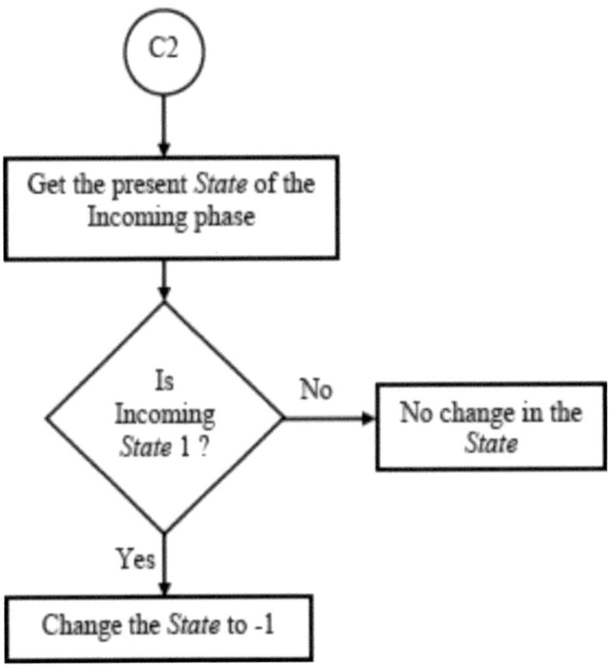
C2
Get the present State of the Incoming phase
Is Incoming State 1 ?
No
No change in the State
Yes
Change the State to -1

http://dx.doi.org/10.5772/intechopen.69280

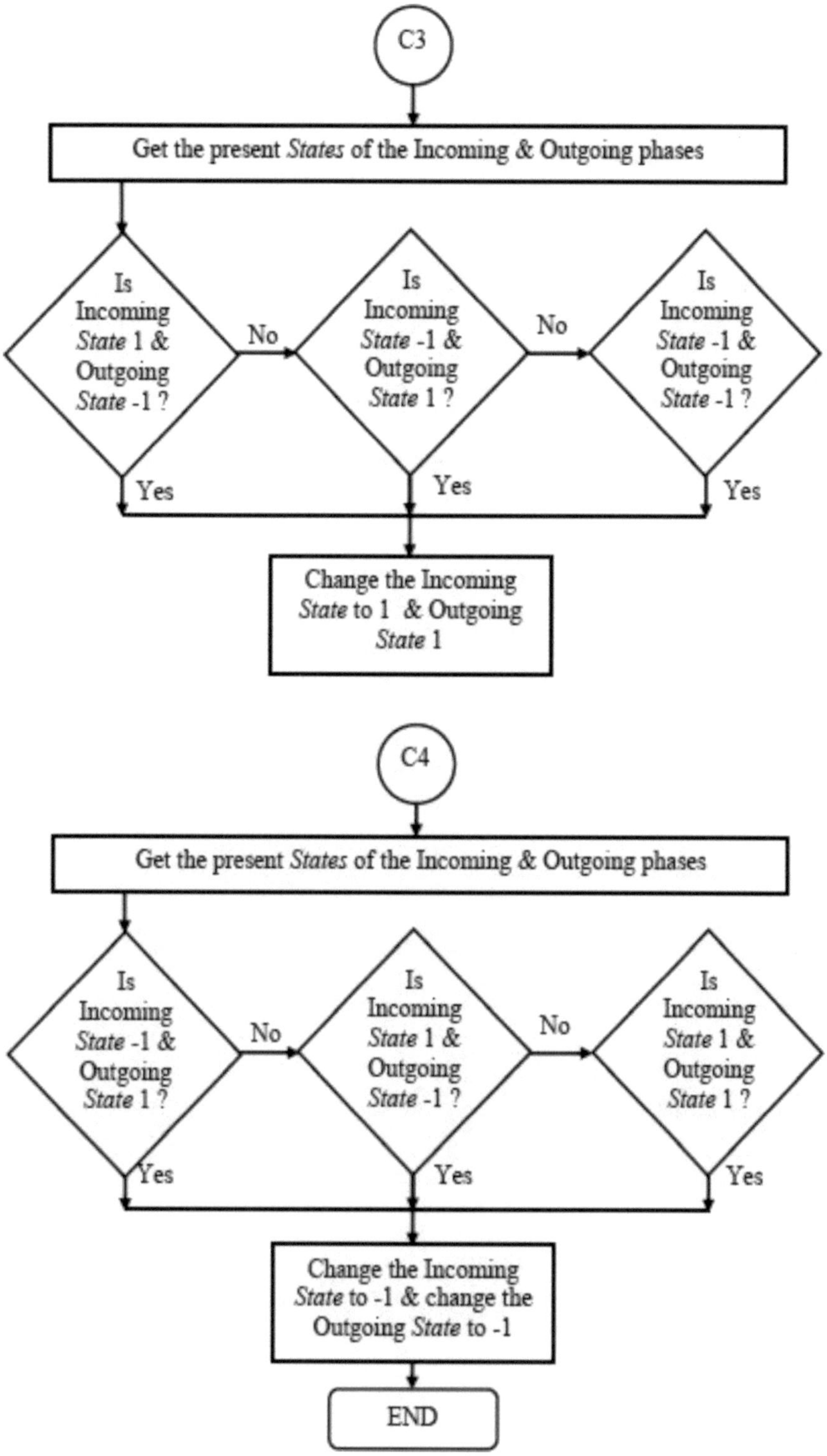

Figure 3. (a–d) Flow chart of DITC.

4. Simulation and analysis of DITC

The complete model of the four-phase 8/6 SRM with DITC controller is shown in **Figure 4(a)**. The specifications of the SRM are given in Appendix A. The model consists of electrical system, mechanical system, position sensing, asymmetrical converter, and DITC controller blocks. DITC program is written in the embedded function blocks described in **Figure 3**. **Figure 4(b)** shows the internal model of SRM. The performance of the DITC-based SRM drive is analyzed for a fan-type load of 8 Nm and at a reference speed of 800 rpm. The change in the switching frequency of the device with change in hysteresis band is shown in **Table 1**. The typical frequency of the device is from 5 to 20 kHz. For 5% band of torque hysteresis, the switching frequency is less than 15 kHz, which is the safe operating frequency of the device. Thus, 5% torque hysteresis band can be selected for this drive, because a further decrease in the hysteresis band increases the switching frequency which is beyond the normal operating frequency of the device. It is observed that at lower hysteresis bands, switching frequencies of the device are higher, which increases the switching losses and reduces the efficiency [14].

Total torque of the motor in steady state is shown in **Figure 5(a)**. It is calculated that the torque band is 0.432 Nm as against the set band of 0.40 Nm, while the set band is 0.40 Nm and the torque ripple is 5.41%. **Figure 5(b) (c)** show the voltage applied across two phases, Ph1 and Ph2, respectively. Ph1 is the outgoing phase and Ph2 is the incoming phase. The voltage in each phase changes between +120 and −120 V, in order to maintain total torque same as the load torque. The torque sharing between two consecutive phases is shown in **Figure 5(d)**. During

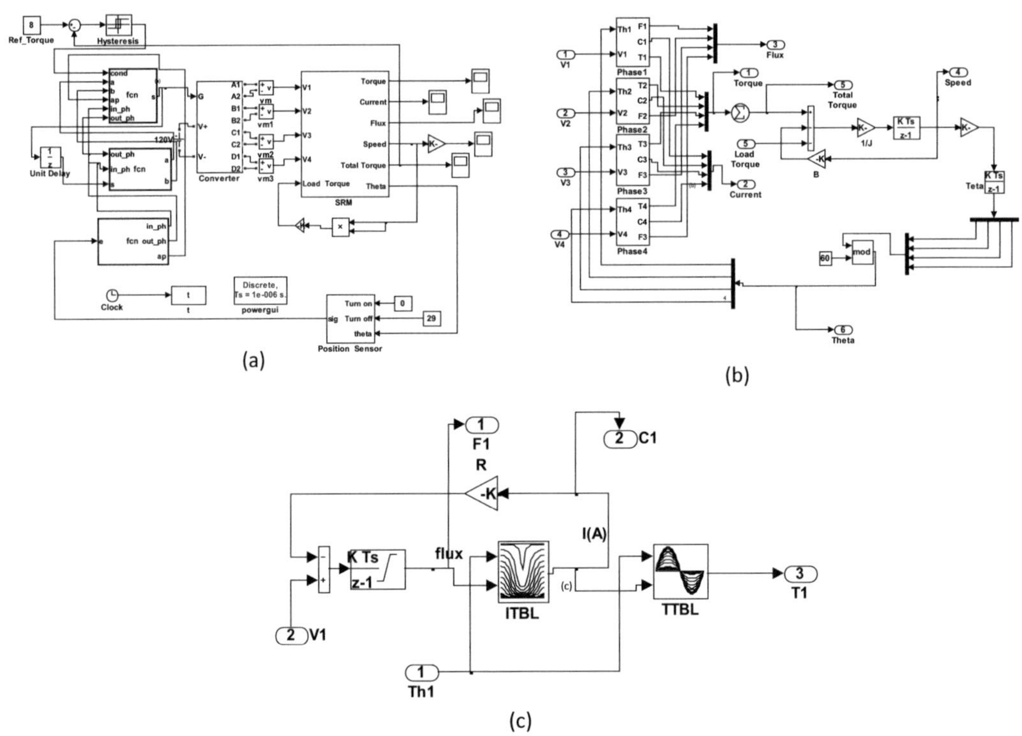

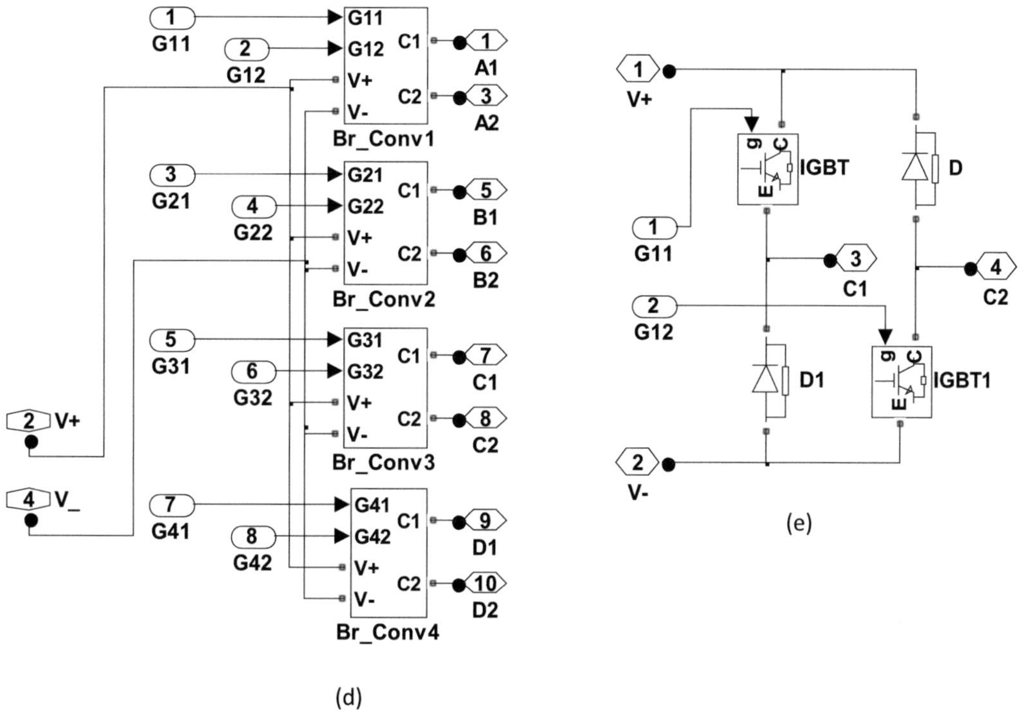

Figure 4. (a) Simulation diagram of SRM drive with DITC, (b) internal block of SRM, (c) internal model of phase 1, (d) asymmetrical converter, and (e) one leg of the converter.

Torque hysteresis band (%)	Switching frequency (f_s) (kHz)
10	6.22
8	8.89
5	13.99
4	16.18
3	21.54

Table 1. Variation of switching frequency with torque hysteresis band.

commutation, it is observed that outgoing torque is decreasing and incoming torque is increasing, but the total torque is kept constant by changing the *States* between 1 and −1. **Figure 6** shows the expanded view of the **Figure 5**.

Torque of all the four phases in the steady state is shown in **Figure 7(a)**, while the current in all the phases is shown in **Figure 7(b)**. The maximum current and average current in each phase are 15.50 and 4.62 A, respectively. **Figure 7(c)** shows the total torque response of the motor. Thus, the torque ripple is minimized in the steady state and during acceleration period. **Figure 7(d)** shows the load torque. **Figure 7(e)** shows the speed response. The steady state speed is maintained constant at 798 rpm. The settling time of the speed is 0.32 s.

The performance of the DITC-based drive is also analyzed for a constant torque load of 8 Nm and at a reference speed of 800 rpm. The simulation diagram is shown in **Figure 8**. The input to the PI controller is the error between the actual speed and the command speed and the output

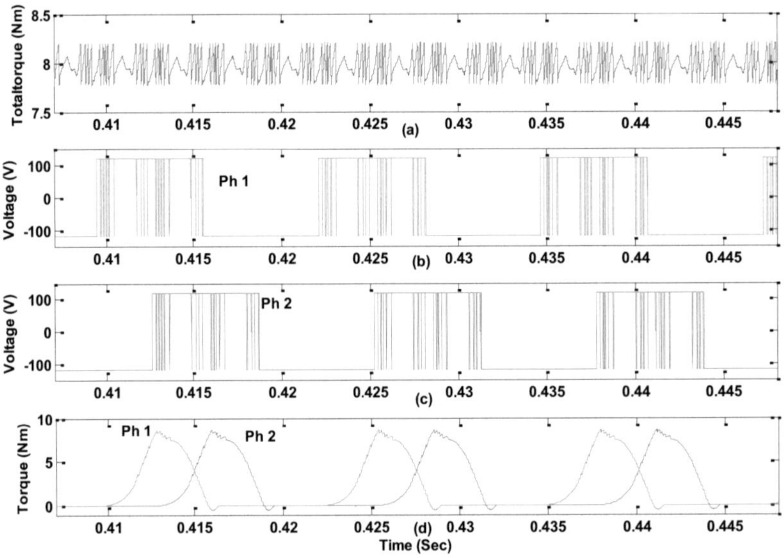

Figure 5. DITC with fan-type load (a), total torque (b), (c) phase voltages, and (d) phase torques.

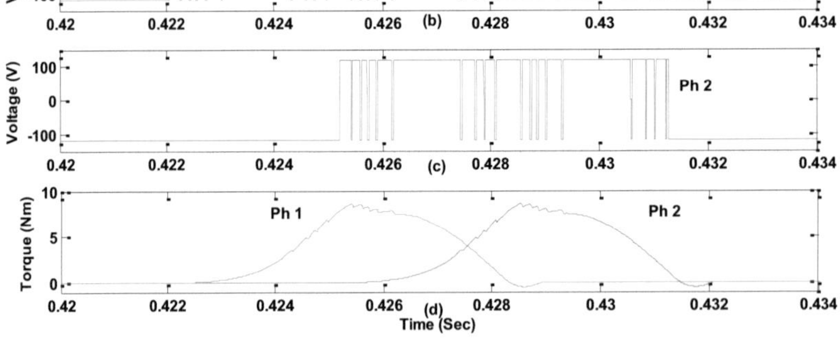

Figure 6. Expanded view of Figure 5.

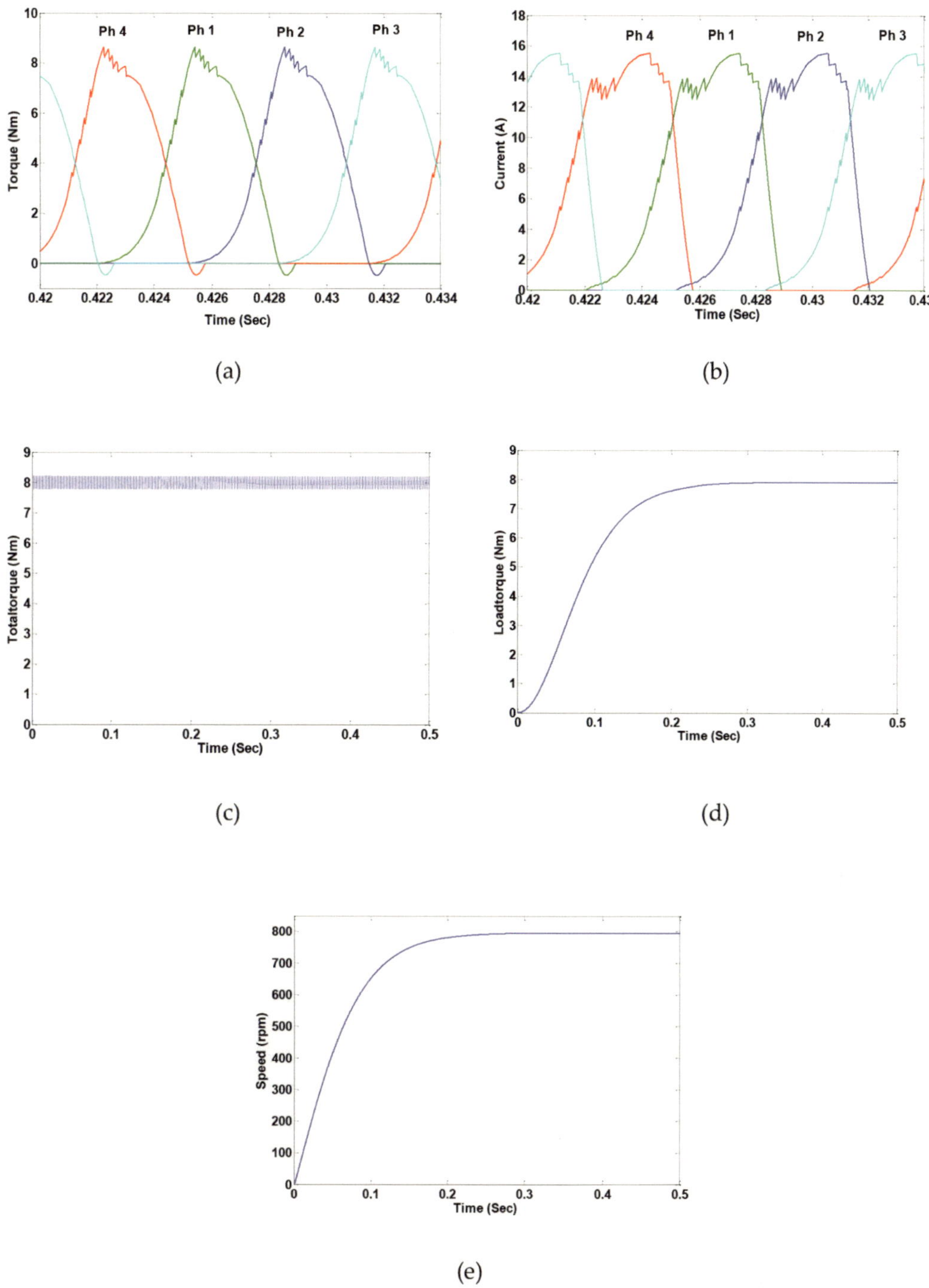

Figure 7. DITC with fan-type load (a), phase torques (b), phase currents (c), total torque (d), load torque (e) speed.

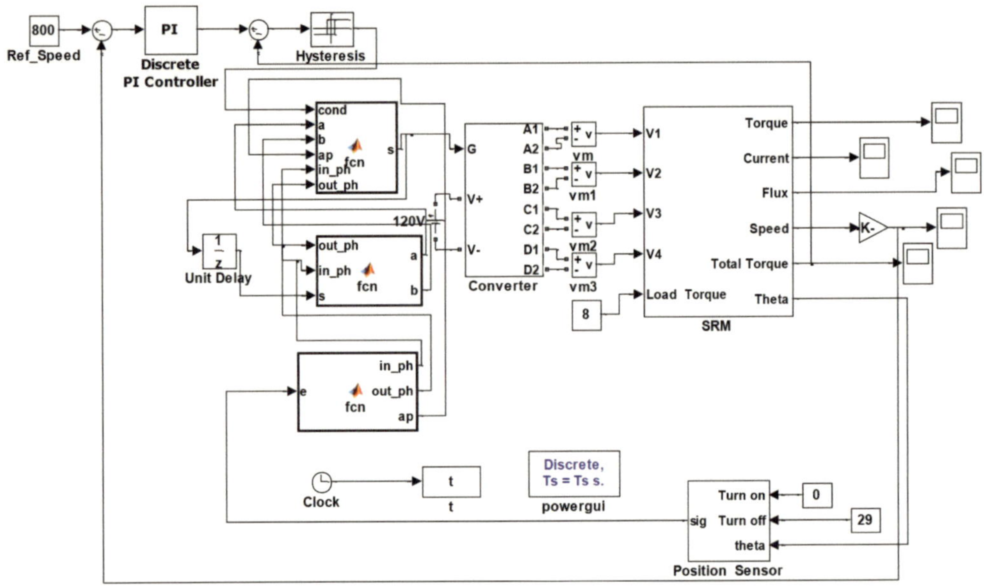

Figure 8. Simulation diagram of SRM drive with direct instantaneous torque controller for constant torque load.

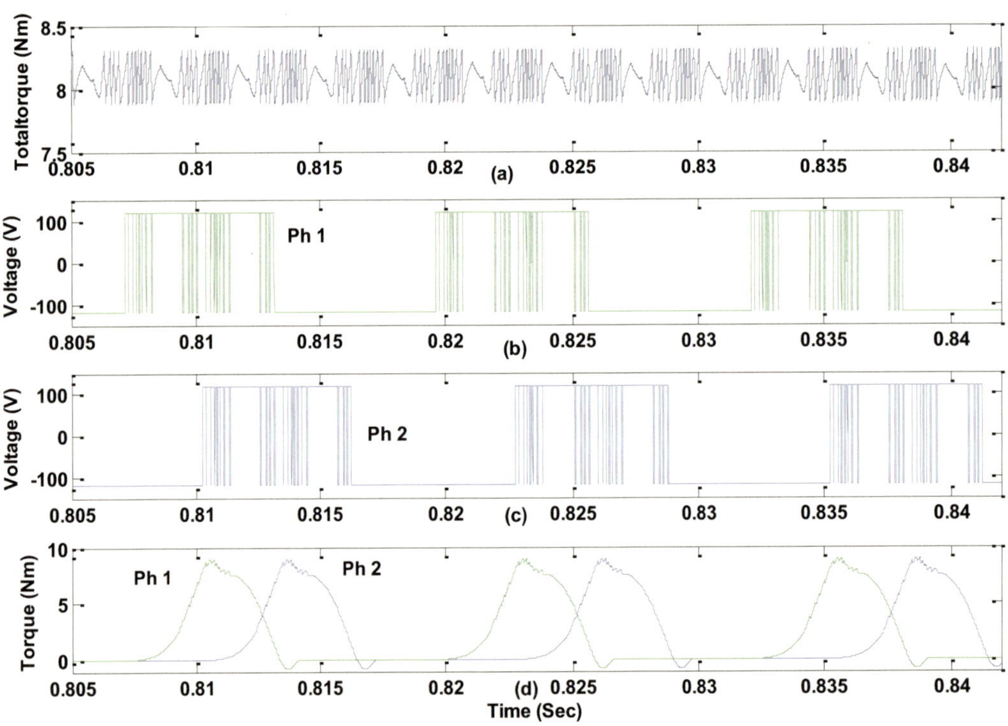

Figure 9. DITC with constant torque load (a), total torque (b), (c) phase voltages, and (d) phase torques.

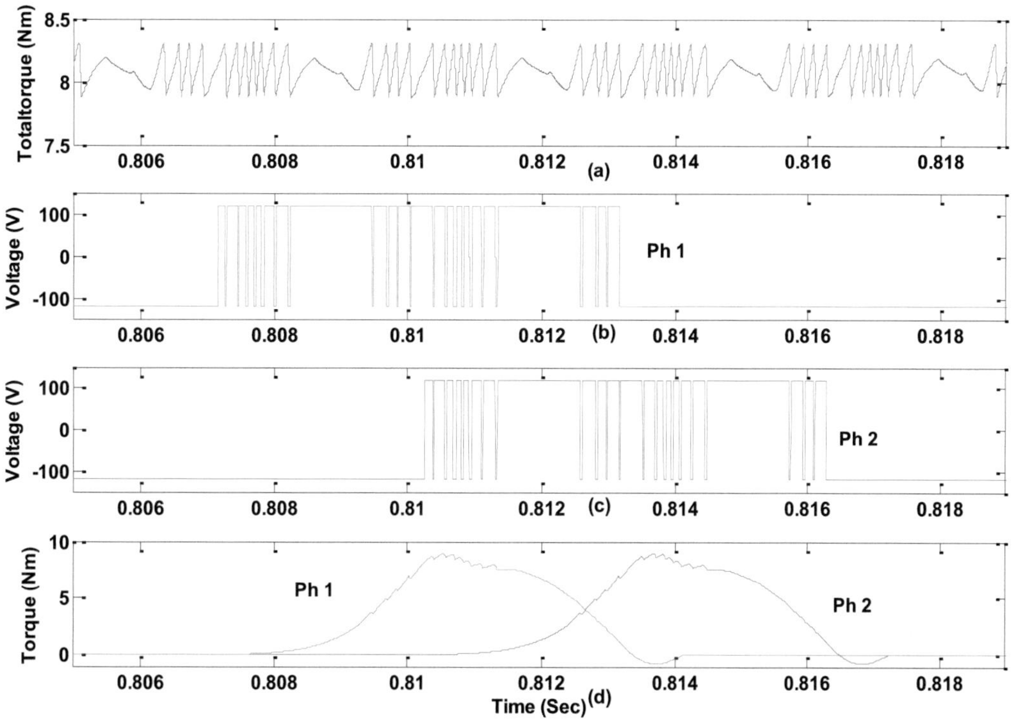

Figure 10. Expanded view of Figure 9.

is the command torque. Whenever the speed is less than the command speed, the output of the PI controller is 9 Nm. The output of PI controller is equal to the load torque when the speed reaches the reference value.

Simulation waveforms of the DITC-based drive for constant torque load are shown from **Figures 9(a)–(d)**, **10**, and **11(a)–(e)**. It is observed that the torque is maintained within a band of 0.43 Nm as against the set band of 0.40 Nm. The torque ripple is only 5.38% in steady state. The maximum current and average current in each phase are 17.59 and 5.00 A, respectively. Thus, the torque ripple is minimized in the steady state and during acceleration period. The settling time of the speed is 0.713 s. The variation of peak current, average current, torque ripple, and speed settling time for DITC-based SRM drive is given in **Table 2**.

From the **Table 2**, it is observed that the peak current, average current, and settling time is higher for constant torque load as compared to fan load. There is no significant variation in the torque ripple for both types of loads. Switched reluctance motor drives with sophisticated direct instantaneous torque control can minimize the torque ripples and increases the efficiency of the drive when operated at the required torque band and thus can be included in the future energy technologies.

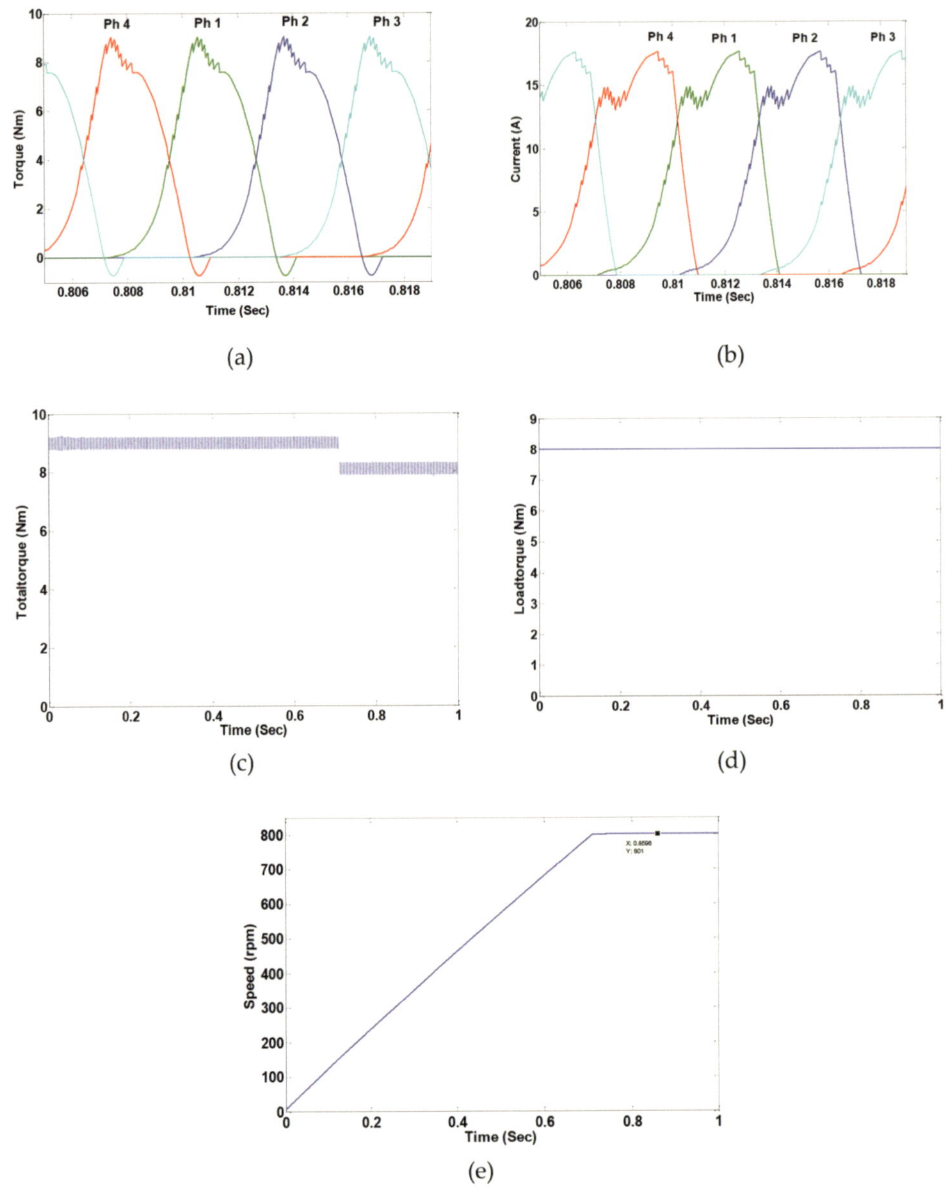

Figure 11. DITC with constant torque load (a), phase torques (b), phase currents (c), total torque (d), and load torque (e) speed.

Type of load	Peak current (A)	Average current (A)	% Torque ripple	Settling time (s)
Fan-type load	15.50	4.62	5.41	0.32
Constant torque load	17.59	5.25	5.38	0.71

Table 2. Comparative analysis for fan-type load and constant torque load with DITC.

http://dx.doi.org/10.5772/intechopen.69280

5. Conclusion

A more sophisticated control technique named direct instantaneous torque control is presented in this chapter. In this control strategy, the torque is maintained within a set hysteresis band by changing the switching *States* of the phases. The torque ripple variation with switching frequency has been analyzed with DITC controller under accelerating and steady state conditions.

The DITC controlled drive is simulated in MATLAB/SIMULINK environment with fan load and constant torque and is observed that the controller maintains the total torque within its set band during steady state and also while accelerating. The switching frequency under such conditions is 13.99 kHz. Thus, it can be concluded that, unlike conventional techniques, DITC control does not require any current profiles or torque sharing functions.

A comparative analysis of DITC-based drive with constant torque load and fan-type load is made. It is observed that the peak current and average current are higher for constant torque load compared to fan-type load. The torque ripple is almost same for both types of loads. The speed settling time for constant torque load is more than twice the value obtained with fan-type load.

Appendix A

Voltage	120 V DC
Maximum current	30 A
Rated speed	1500 rpm
Maximum flux	0.3 Wb
Aligned inductance	110 mH
Unaligned inductance	10 mH
Resistance	0.3 Ω
Stator poles	8
Rotor poles	6

Author details

Srinivas Pratapgiri

Address all correspondence to: srinivasp.eedou@gmail.com

Department of Electrical Engineering, University College of Engineering, Osmania University, Hyderabad, India

References

[1] Miller TJE. Switched Reluctance Motors and Their Control. Magna Physics Publishing & Oxford University Press; 1993

[2] Miller TJE. Electronic Control of Switched Reluctance Machines. Oxford & Boston: Newnes Press; 2001

[3] Krishnan R. Switched Reluctance Motor Drives: Modeling, Simulation, Analysis, Design, and Applications. CRC Press; 2001

[4] Emadi A. Energy-Efficient Electric Motors. 3rd ed. Marcel Dekker; 2005

[5] De Doncker R, Pulle DWJ, Veltman A. Advanced Electrical Drives: Analysis, Modeling and Control. Springer; 2011

[6] Wenzhe Lu. Modeling and control of switched reluctance machines for Electro-Mechanical braking system [thesis]. The Ohio State University; 2005

[7] Wang X. Modeling and implementation of controller for switched reluctance motor with AC small signal model [thesis]. Virginia Polytechnic Institute and State University; 2001

[8] Cheok AD, Hoon PH. A new torque control method for switched reluctance motor drives. In: 26th Annual Conference of the IEEE Industrial Electronics Society; 22-28 October 2000; Nagoya, Japan. IEEE; 2000. pp. 387–392

[9] Cheok AD, Fukuda Y. A new torque and flux control method for switched reluctance motor drives. IEEE Transactions on Power Electronics. 2002;**17**(4):543–557

[10] Inderka RB, De Doncker RW. DITC-Direct instantaneous torque control of switched reluctance drives. IEEE Transactions on Industry Applications. 2003;**39**(4)

[11] Fuengwarodsakul NH, Menne M, Inderka RB, De Doncker RW. High_dynamic Four-Quadrant switched reluctance drive based on DITC. IEEE Transactions on Industry Applications. 2005;**41**(5)

[12] Chancharoensook P. Direct instantaneous torque control of a Four-Phase switched reluctance motor. In: International Conference on Power Electronics and Drive Systems; 2-5 November 2009; Taipei, Taiwan. IEEE; 2009. pp. 770–777

[13] Inderka RB, Krehenbrink M, De Doncker RW. On-line Estimation of Instantaneous Torque of Switched Reluctance Machines. In: International Symposium on Industrial Electronics; 4-8 December 2000; Cholula, Puebla, Mexico. IEEE; 2000

[14] Srinivas P, Prasad PVN. Torque ripple minimization of 4 phase 8/6 switched reluctance motor drive with direct instantaneous torque control. International Journal on Electrical Engineering and Informatics. 2011;**3**(4):488–497

Chapter 4

Design, Power Electronics and Torque Control of Switched Reluctance Machines

Mircea Ruba and Petre Dorel Teodosescu

Additional information is available at the end of the chapter

http://dx.doi.org/10.5772/intechopen.69270

Abstract

In the last decade, increased tendency in the field of automotive industry was focused on the development of highly efficient and low-cost electric propulsion systems to replace the existing internal combustion solutions. The aim is to reduce the pollution due to carbon dioxide emissions into the air. Several electric machine topologies with their power electronics, control and supply units are continuously in the development process to reach the desired goal. One such machine is the switched reluctance machine (SRM), reaching increased power density, low cost and possibility of continuous operation despite fault occurrence. Designing the machine, choosing its power electronics and controlling the machine to diminish the negative effect of the torque ripples are key points in reaching the proper propulsion system. The main topics presented in detail in this chapter are managing the reader's skills with an analytic design breviary, presenting the machine's control strategies for instantaneous torque linearization and finally, showing a power converter topology with increased performances in low voltage applications. To be more close to such an application, the exampled machine is developed for a light electric vehicle for people with physical disabilities. Operational skills of the machine will be validated based on complex simulations.

Keywords: switched reluctance machine (SRM), design breviary, power electronics, torque smoothening, light electric vehicle

1. Introduction

Nowadays, research activity in the field of automotive industry receives a strong influence due to necessity to reduce the emissions of polluting gasses in the atmosphere. Hence, replacing the classical internal combustion engines with electric propulsion systems that are non-polluting became a hot topic in research labs all over the world [1]. Different electrical machine structures, their power electronics and supply units are continuously developed and tested, aiming the goal of high efficiency at as low costs as possible. Induction machines and permanent magnet machines are now implemented on Tesla and Toyota electric vehicles. However, their design and building costs are higher compared with other machine topologies, such as the switched reluctance machine (SRM). The latter has the main drawback of more complex power electronics and complex control. However, with the advance of the semiconductors and processors, it became possible to develop reduced price electronics and their applied control for SRMs.

Another feature of the SRM is the ability to continuously operate even in faulted conditions and if added to the original topology, structural modifications one can reach a highly fault tolerant propulsion machine [2].

Designing an SR machine is not too complicated, however, it is important to establish during sizing process, proper flux density values in the magnetic core. Too low values have the outcome of an unsaturated machine and by this, poor power density and too high values will limit the developed power and cause core heating [3]. Proper sizing of the air-gap is having huge performance influence; hence, a compromise must be considered between low values and building costs that are increasing drastically for values smaller than 0.5 mm.

Choosing and designing the proper architecture for the power electronics for an SR machine is crucial from the point of view regarding costs and ability to perform torque linearization control. If no such dedicated control is required, only classical hysteresis one is engaged, there are simple power converters that can be used with minimal number of transistors. However, if one desires to develop and use torque smoothening procedures, topologies that allow these are mandatory to be used, based on complete or half-H bridge designs, with independent switch control. Moreover, the driver that turns on/off the power switches must be able to maintain the state of the switch for an unlimited time [4].

A serious drawback of the SRM is the increased torque ripples that are caused due to the working principle of the machine to switch the current from one phase to the next one that encounters the lowest magnetic reluctance. By this, the rotor moves from unaligned to aligned rotor to stator poles, the movement being characterized by a sudden instantaneous torque variation. Part of the torque ripple minimization can be handled during the design phase [5], shaping correctly the rotor poles function of the stator ones. However, this is limited up to an extent that is still considered too much for automotive applications. The main method for decreasing as much as possible the torque ripples that create mechanical stress, noise and vibrations is to engage direct instantaneous torque control (DITC) of current profiling based on torque sharing functions (TSF). For both, as mentioned in the previous paragraph, it is necessary to have certain electronics that allow their operation. Following the details presented in this chapter, one can see that the toque characteristic of the SRM can reach a shape just like that of an induction or synchronous motor, yet using a much cheaper and more simple machine structure.

To be more comprehensive when designing such a machine, it is more transparent if one considers a specific application. For this case, the application will be a light electric vehicle designed for people with physical disabilities. Based on an existing DC machine mounted on such a vehicle, the main parameters that will be the start-up for the design process are: the required output power (P_{2N}) 1.2 kW, the supply voltage (U_N) of 24 V, shaft speed (n_N) of 3400 rpm and electromagnetic required torque (T) of 3.4 Nm. Besides these considerations, it is mandatory that the machine needs to fit in the place of the existing DC machine. Hence, the maximum dimensions allowed are as follows: for the outer diameter 115 mm and for the active stack length 150 mm.

2. Design of the switched reluctance machine

Based on the specifications detailed in the introduction part, one can start sizing the SRM. Before that the stator (Q_S) and rotor (Q_R) pole numbers must be imposed. Usually, three-phase SRMs are cheaper both in electronics and machine building, but encounter high torque ripples, while five- or six-phase machine with low torque ripples reach increased development costs. Hence, the best compromise is to develop a four-phase machine, with a Q_S/Q_R ratio of 8/6 [6]. Another parameter that must be imposed is the flux density toward the air-gap (B_{gmax}) at 1.9 T. The current computed function of the supply voltage and the machine's requested power, considering an efficiency of 0.65 (low power SRMs have quite poor efficiency) is $I = 80$ A. The design process is an iterative one, as this will be explained later. Hence, the air-gap was set to a low value of $g = 0.1$ mm, this is due to the dimension limitations of the outer machine diameter.

The most influential parameter of the machine is the mean diameter [7] (D_g).

$$D_g = \sqrt[3]{\frac{P_{2N} \cdot Q_s \cdot k_\sigma}{Q_R \cdot \pi^2 \cdot k_L \cdot \frac{n_N}{60} \cdot B_{gmax} \cdot \left(1 - \frac{1}{K_{cr}}\right) \cdot A_S}} \tag{1}$$

In Eq. (1), k_σ and k_L coefficients are the leakage flux factors, chosen between 0.75 and 0.95, respectively, the aspect factor, which can be calculated from the rotor pole number using Eq. (2).

$$k_L = \frac{\pi}{2} \cdot \frac{1}{\sqrt[3]{Q_R}} \tag{2}$$

It should be mentioned that for this particular design, as the active stack length is given in the specifications, one does not need to compute it any more. The term A_s represents the electrical loading that is chosen in the range 25,000–100,000 A/m, where higher values correspond to smaller dimensions [8]. Carter's factor (K_{CR}) considers the flux path's distortion due to the shape of the salient poles. Its value ranges between 1.4 and 2. The ratio of the mean diameter with respect to the aspect factor will give an estimate of the active stack length of the machine (l_S), but as mentioned for this design, this is a known value.

The stator and rotor pole pitch is computed as ratio between the mean diameter and the number of the poles, using Eq. (3).

$$\tau_{S,R} = \pi \cdot \frac{D_g}{Q_{S,R}} \tag{3}$$

The width of the stator and rotor poles (b_{pS} and b_{pR}) can be calculated by using the pole pitch values, considered about 0.1–1.3 of it. The values chosen for stator and rotor were, respectively, 0.8 and 1.15 to reach as low torque ripples as possible [6]. Using these values, the yokes of the stator and rotor can be computed considered in the range of 0.5–1 of the pole width. Smaller values will reach proper saturation in the magnetic cores [5, 9] which will help to extract the energy from the coils when the phase is turned off. However, too low values will saturate the core too much resulting in the overheating of machine.

The slot openings are used to determine function of the stator and rotor pole widths and the pole pitch values computed with Eq. (3).

$$b_{cS,cR} = \tau_{S,R} - b_{pS,pR} \tag{4}$$

Using a catalogue value for the shaft (d_{ax}) one can now finalize the sizing process that regards the SRM's rotor. The rotor pole height, meaning the dimension from the airgap to the rotor yoke, and the inner rotor diameter will be computed using Eqs. (5) and (6), both function of the rotor yoke height (h_{jR}).

$$h_{pR} = \frac{D_g - g}{2} - h_{jR} - \frac{d_{ax}}{2} \tag{5}$$

$$D_{iR} = D_g - g - h_{jR} - h_{pR} \tag{6}$$

To finalize the sizing process with regard to the stator; firstly, it is necessary to size the coils of the machine, because these influence the height of the stator poles. This process starts from the magnetomotive force (mmf). There are several methods to compute it, but one efficient and simple way is to take advantage of the known parameters, such as the air-gap length (g) and its flux density value (B_{gmax}), the saturation factor, the flux leakage factor (k_ρ) and the relative permeability of the air (μ_0).

$$\Theta = \frac{g \cdot k_{sat} \cdot B_{gmax}}{k_\rho \cdot \mu_0} \tag{7}$$

Now a round value of the number of turns can be established as a function of the magnetomotive force and the phase current (I) of the SRM.

$$N_f = \mathrm{round}\left(\frac{\Theta}{I}\right) \tag{8}$$

Next, sizing the cross-section of the wire has handled function of the current density (J_c) in the range of 2.5–8 A/cm^2.

$$S_{cond} = \frac{I_{rms}}{J_c} \tag{9}$$

In Eq. (9), it is seen the rms value of the current (I_{rms}) is used instead of its rated value. The reason is that the current is switched from one phase to another, and the machine has four phases; one phase will be energized for only ¼ of the entire period. Hence, the rms current will be computed as the function of the number of phases (N_{phase}) like in AC supply systems, based on Eq. (10).

$$I_{\text{rms}} = \frac{I}{\sqrt{N_{\text{phase}}}} \tag{10}$$

The used wire diameter can be computed as the function of the cross-section in Eq. (9). This value will be used for arranging the turns inside the stator slots, as it will be presented as follows. The total number of turns is placed in several layers. Each layer will have a certain number of turns ($N_{\text{sp\_strat}}$), computed function of the slot opening (b_{cS}), its insulation (g_{iz}) and the actual wire diameter.

$$N_{\text{sp\_strat}} = \text{round}\left(\frac{l_{\text{bob}} - 2 \cdot g_{\text{iz}}}{1.05 \cdot d}\right) \tag{11}$$

The number of layers (n_{strat}) is found by dividing the total number of turns by the number of turns in a layer. Now, the height of the coil can be used to find the function of the number of layers, the wire diameter and the isolation of each wire ($g_{\text{iz\_strat}}$), using Eq. (12).

$$h_{\text{bob}} = n_{\text{strat}} \cdot (d + g_{\text{iz\_strat}}) + 2 \cdot g_{\text{iz}} \tag{12}$$

The area inside the stator slot occupied by the coil is used to compute function of the number of turns, the cross-section of the wire and a factor (k_{u}) ranged between 0.5 and 0.85 that considers the wire surface imperfections.

$$A_{\text{bob}} = N_{\text{f}} \cdot \frac{S_{\text{cond}}}{k_{\text{u}}} \tag{13}$$

The term A_{bob} is used for only half of the stator slot. In one slot, two such areas need to fit as two phases sharing the same slot. Hence, the total slot area will be computed using Eq. (14).

$$A_{\text{c}} = 1.1 \cdot 2 \cdot A_{\text{bob}} \tag{14}$$

The term '1.1' is added as a safety caution because there are other imperfections of the coils that cannot be taken into calculation every time. Having these dimensions fixed, one can compute the height of the coil using Eq. (15).

$$h_{\text{cS}} = \text{round}\left(\frac{A_{\text{c}}}{b_{\text{cS}}}\right) \tag{15}$$

Usually for simplifying the cutting process, round values are imposed. At this point, it is easy to find the height of the stator pole (h_{pS}), adding to the height of the coil the height of a nonmagnetic displacer (h_{lim}) used to fix the coil into the slot.

$$h_{pS} = h_{cS} + h_{lim} \tag{16}$$

At this point, having all the required dimensions, it is possible to compute the outer diameter of the machine, for this project, to check if this does not exceed the imposed value.

$$D_M = D_g + g + 2 \cdot h_{pS} + 2 \cdot h_{jS} \tag{17}$$

To be able to compute the resistance of one phase winding, it is necessary to know the wire length function of the machine dimensions.

$$l_{inf} = 2 \cdot N_f \cdot (l_S + h_{jS}) \tag{18}$$

Finally, the resistance is given by Eq. (19) considering the cross-section and the material properties.

$$R = \rho_{Cu} \cdot \frac{l_{inf}}{S_{cond}} \tag{19}$$

2.1. Analytic calculation of losses and torque in SR machines

To calculate the efficiency of the newly designed SR machine, different methods for losses approximation can be used [7, 10]. Preliminary, it is mandatory to compute the frequency of the flux density variation in the magnetic core, both for the stator and the rotor.

$$\begin{aligned} f_{jS} &= \frac{Q_S}{2} \cdot \frac{n_N}{60} \\ f_{jR} &= \frac{Q_R}{2} \cdot \frac{n_N}{60} \end{aligned} \tag{20}$$

The specific losses in the machine's core, computed for standard values measured for 50 Hz and 1 T are calculated using Eq. (21), where B_{jS} and B_{jR} are the stator and rotor yoke flux densities.

$$\begin{aligned} p_{FeS} &= p_{Fesp} \cdot B_{jS} \cdot f_{jS} \\ p_{FeR} &= p_{Fesp} \cdot B_{jR} \cdot f_{jR} \end{aligned} \tag{21}$$

To calculate the losses in the machine and function of its dimensions, one needs to compute the weight of the assemblies of the core function of the used material's properties.

$$\begin{aligned} G_{FejS} &= \pi \cdot \left[\frac{D_M^2}{4} - \left(\frac{D_M}{2} - h_{jS}^2 \right)^2 \right] \cdot l_S \cdot \rho_{Fe} \\ G_{FejR} &= \pi \cdot \left[\left(h_{jR} + \tfrac{d_{ax}}{2} \right)^2 - \left(\tfrac{d_{ax}}{2} \right)^2 \right] \cdot l_S \cdot \rho_{Fe} \\ G_{FepS} &= Q_S \cdot b_{pS} \cdot h_{pS} \cdot l_S \cdot \rho_{Fe} \\ G_{FepR} &= Q_R \cdot b_{pR} \cdot h_{pR} \cdot l_S \cdot \rho_{Fe} \end{aligned} \tag{22}$$

The losses in the winding have computed function of the internal resistance and the phase current.

http://dx.doi.org/10.5772/intechopen.69270

$$P_{\mathrm{j}} = R \cdot I^2 \tag{23}$$

The mechanical losses are estimated approximately 0.5% of the machine's output power.

$$P_{\mathrm{M}} = 0.005 \cdot P_{\mathrm{2N}} \tag{24}$$

Finally, the total losses of the machine will be the sum of the above calculated ones:

$$P_{\mathrm{T}} = P_{\mathrm{j}} + P_{\mathrm{Fe}} + P_{\mathrm{M}} \tag{25}$$

Hence the efficiency will be

$$\eta_{\mathrm{SRM}} = \frac{P_{\mathrm{out}}}{P_{\mathrm{out}} + P_{\mathrm{T}}} \tag{26}$$

The developed torque can be used to compute function of the mmf created in the machine and its main dimensions using Eq. (29).

$$T_{\mathrm{SRM}} = 2 \cdot (N_{\mathrm{f}} \cdot I^2) \cdot \frac{D_{\mathrm{g}}}{2} \cdot \mu_0 \cdot \frac{l_{\mathrm{S}}}{2g} \tag{27}$$

The term $2g$ (twice the air-gap) in Eq. (29) stands because always two diametrically opposed poles contribute to the torque development, hence, the air-gap length is considered double along the flux path.

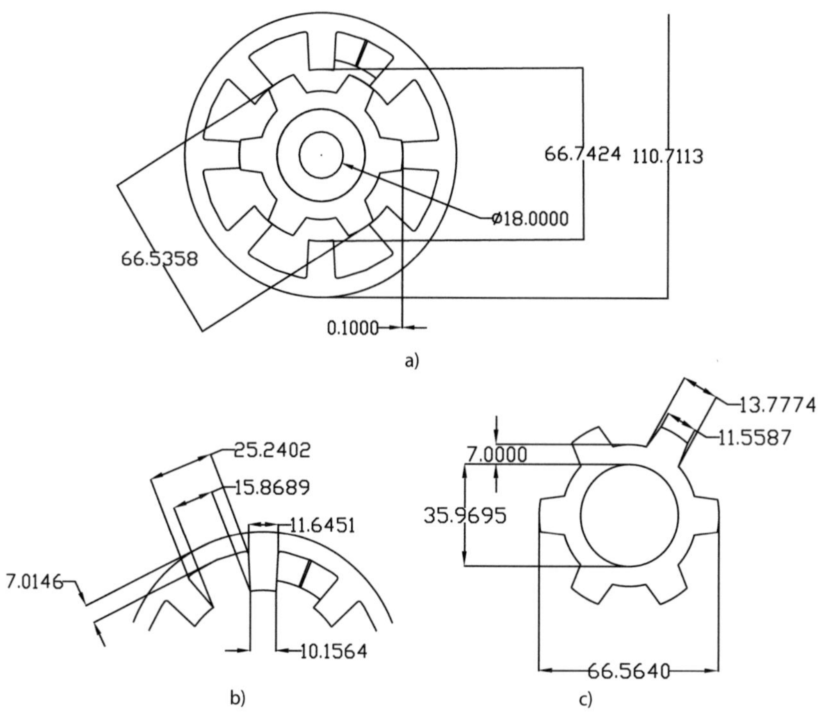

Figure 1. The resulted SRM dimensions: (a) the entire machine, (b) details of the stator and (c) details of the rotor.

Starting from the requirements detailed at the beginning of the chapter and using the above presented breviary, an SRM designed for light electric vehicle was obtained with the main dimensions which are depicted in **Figure 1**.

To certify that the machine meets the requirements of developing at 80 A, a torque of 3.4 Nm at 3400 rpm, in finite element analysis (FEA) model was created in Cedrat Flux 2D software. The current in the windings was handled using hysteresis controller referenced at 80 A as depicted in **Figure 2a**. At rated current the mean torque reaches the value of 3.4 Nm, but, despite attempts to reduce the torque ripple by design, these are still quite high. In such cases, the SRM cannot be used for electric propulsion systems as those ripples create high noise and vibration in the mechanical transmission and the car's body itself.

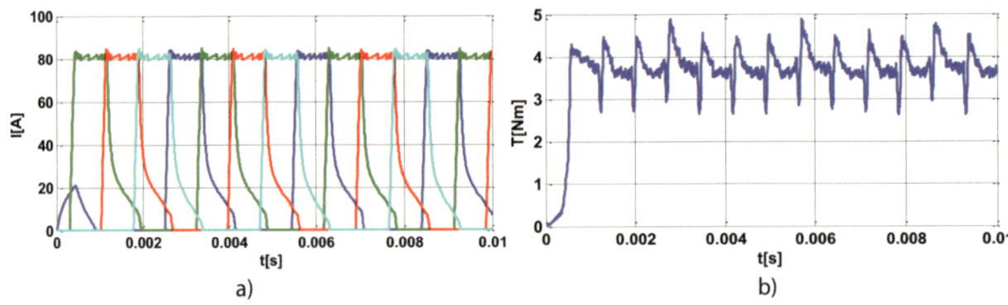

Figure 2. The simulation results of the designed SRM: (a) the phase currents and (b) the developed torque.

3. Torque linearization control strategies for the SRM

The results shown in **Figure 2** indicate that the operational skills of the SRM, but more, it is also proved that the torque ripples are too high compared to the requirements of an electric vehicle propulsion unit. The structure recorded a success up to an extent. However, the torque ripples need to be further reduce to fit the machine in the EV requirements. For this purpose, special control procedures are engaged, such as direct instantaneous torque control (DITC) [11, 12] or current profiling based on torque sharing functions (TSF) [13].

Before detailing each of the above-mentioned methods, some requirements regarding their implementation must be highlighted. Besides a good knowledge about the parameters of the machine, of the power converter and of the controller's sampling frequency, each of the two methods is based on inserting into the control model look-up-tables (LUT). For DITC, the LUT must contain information of the variation of the torque versus current and rotor position, as depicted in **Figure 3a**. Usually the content of this LUT is fetched from the FEA model of the machine. If the laboratory facility permits it, it is better to record this data from experimental measurements. However, as its name mentions it, DITC is an instantaneous torque control; hence, at each computation sample, the controller must have precise information of the torque values. This information can be extracted from the LUT, knowing precisely the shaft position and the measured phase current.

The second control strategy, the current profiling based on TSF, requires a reversed structure of the LUT depicted in **Figure 3a**, having the variation of the current versus torque and rotor

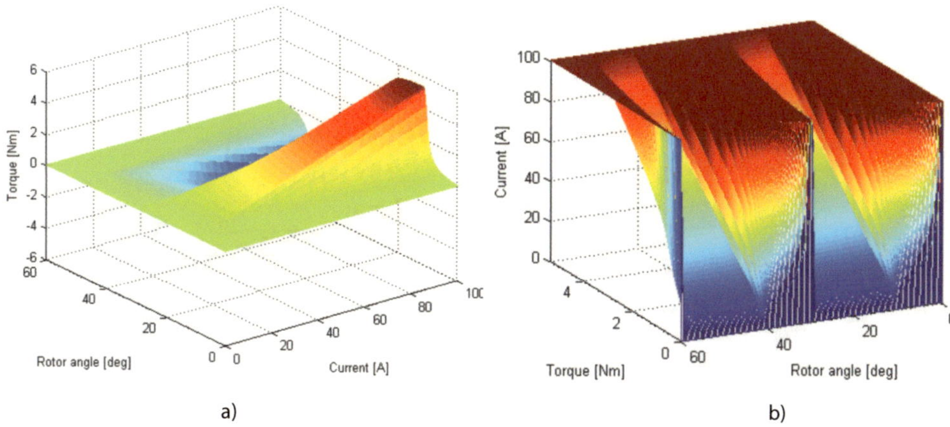

Figure 3. The look-up-table used for the DITC (a) and TSF (b).

position. The data can be obtained proceeding for a reversed interpolation of the torque versus current and rotor position, with respect to the phase current. This LUT is depicted in **Figure 3b**.

3.1. Direct instantaneous torque control (DITC)

The DITC (see **Figure 4a**) method [14] invokes a procedure of torque smoothening based on the control of instantaneous torque developed by the machine, using a hysteresis band. The main advantage of this procedure is that it does not require any PI (proportional integral) or PID (proportional integral derivate). The actual shape of the torque is regulated based on a double-layered hysteresis band. The comparison of the torque with the hysteresis band returns directly the gate signals for the power switches. As the torque is not measured directly from the machine, but it is estimated from the LUT, the setup does not require an instantaneous torque transducer which usually costs too much.

Practically, the DITC is implemented using two hysteresis bands, one larger than the other. The controller divides these into three regions, two of them form the torque reference to the upper and lower extremities and the main one is placed in the middle as depicted in **Figure 4a**.

During single phase conduction, the torque is regulated inside the limits of the main (inner) band (as shown in **Figure 4b**). In **Figure 4c**, with blue the voltage of the outgoing phase is depicted, while with red the voltage of the incoming one is represented. With the same colors, in **Figure 4d**, with blue the torque of the outgoing phase is represented while with red the torque of the incoming one is shown. The main involvement of the DITC is reflected during phase commutation. During phase commutation, the torque is regulated by the incoming phase, on one hand, to maintain it inside the inner hysteresis band, and on the other hand, by the outgoing one that becomes energized just enough to increase the torque when it falls and reaches the lower limit of the outer band. On increasing the torque, this will be re-established inside the inner band. At this point, the current increases close to maximum value, hence the torque tends to increase fast. To compensate this issue, the outgoing phase is again energized but with negative voltage (**Figure 4c**) to force fast-fall of the torque to maintain it in the desired

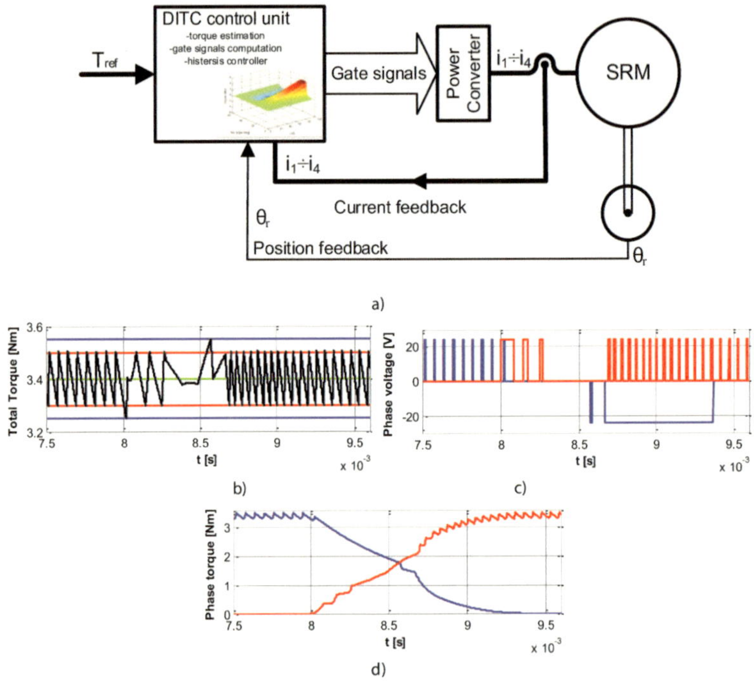

Figure 4. The DITC control scheme (a), the instantaneous torque (b), voltage of adjacent phases and (c) the phase torque (d).

band. From the moment when the incoming phase is energized, despite the times when the torque gets out of the inner hysteresis band, the outgoing phase is kept at zero voltage. Hence, the phase torque is regulated precisely during phase commutation, as seen in **Figure 4c**, time interval when usually the increased ripples appear.

3.2. Torque sharing functions (TSF)

Despite DITC, there are control methods that are more precise based on shaping the current and by this, automatically modifying the torque profile to become close to the linear one. The torque sharing function (TSF) is engaged mainly in the region of phase switching. The outgoing and incoming phase currents are profiled based on specific functions in order to compensate the ripples in the torque characteristic. One important issue that needs to be controlled for this strategy is the overlap angle [15] that needs to be precisely 15 mechanical degrees θ_{ov}:

$$\theta_{ov} \leq \frac{\theta_{rot}}{2} - \theta_{off} \qquad (28)$$

In Eq. (30), θ_{rot} denotes the period of the rotor and θ_{off} the turn off angle; with θ_{on} the turn on angle of the phase is denoted for all the following equations. A general rule for engaging the TSF of the rotor position described in Eq. (31) is valid for all the following analysed cases [16, 17]. As it can be seen there are five levels of the control based on the rotor position. While the machine phase is in non-conducting region, the TSF is null. During the increase and decrease of the current, the slopes are described by functions $f_{inc}(\theta)$ and $f_{dec}(\theta)$. If the region of the phase is in full conduction (neither increase nor decrease of the current), the TSF becomes equal with the reference torque. As it can be seen that the profile of the current is obtained using the LUT data depicted in **Figure 3b**.

$$\mathrm{TSF}(\theta) = \begin{cases} 0 & 0 \le \theta \le \theta_{\mathrm{on}} \\ f_{\mathrm{inc}}(\theta) & \theta_{on} \le \theta \le \theta_{\mathrm{on}} + \theta_{\mathrm{ov}} \\ T_{\mathrm{ref}} & \theta_{\mathrm{on}} + \theta_{\mathrm{ov}} \le \theta \le \theta_{\mathrm{off}} \\ f_{\mathrm{dec}}(\theta) & \theta_{\mathrm{off}} \le \theta \le \theta_{\mathrm{off}} + \theta_{\mathrm{ov}} \\ 0 & \theta_{\mathrm{off}} + \theta_{\mathrm{ov}} \le \theta \le \theta_{\mathrm{p}} \end{cases} \tag{29}$$

In total, there are four different types of TSF named after the mathematical operator that describes them: linear, sinusoidal, exponential and cubic.

The **linear TSF** refers to the fact that the instantaneous torque during phase commutation follows a linear variation with the rotor position. The function that describes this variation is detailed in Eq. (32) for the increasing and decreasing slopes.

$$\begin{aligned} f_{\mathrm{inc}}(\theta) &= \frac{T_{\mathrm{ref}}}{\theta_{\mathrm{ov}}}(\theta - \theta_{\mathrm{on}}) \\ f_{\mathrm{dec}}(\theta) &= T_{\mathrm{ref}} - \frac{T_{\mathrm{ref}}}{\theta_{\mathrm{ov}}}(\theta - \theta_{\mathrm{off}}) \end{aligned} \tag{30}$$

It has to be noted that during phase commutation the incoming and outgoing phases of the machine are both active.

Using **sinusoidal TSF** implies using functions with sinusoidal or co-sinusoidal evolution of the TFS during phase commutation. The model that refers to such variations is detailed in Eq. (33).

$$\begin{aligned} f_{\mathrm{inc}}(\theta) &= 0.5 \cdot T_{\mathrm{ref}} - 0.5 \cdot T_{\mathrm{ref}} \cos\left(\pi/\theta_{\mathrm{ov}} \cdot (\theta - \theta_{\mathrm{on}})\right) \\ f_{\mathrm{dec}}(\theta) &= 0.5 \cdot T_{\mathrm{ref}} + 0.5 \cdot T_{\mathrm{ref}} \cos\left(\pi/\theta_{\mathrm{ov}} \cdot (\theta - \theta_{\mathrm{off}})\right) \end{aligned} \tag{31}$$

In Ref. [11], the functions detailed in Eq. (33) are presented only function of the on and off angles of the phase. Here, in order to improve the functionality, the overlap angle is also introduced.

The third model, the **exponential TSF**, considers the on and off angles, the actual rotor position and the overlap angle too, as detailed in Eq. (34).

$$\begin{aligned} f_{\mathrm{inc}}(\theta) &= T_{\mathrm{ref}}\left[1 - \exp\left(\frac{-(\theta - \theta_{\mathrm{on}})^2}{\theta_{\mathrm{ov}}}\right)\right] \\ f_{\mathrm{dec}}(\theta) &= T_{\mathrm{ref}}\left[\exp\left(\frac{-(\theta - \theta_{\mathrm{off}})^2}{\theta_{\mathrm{ov}}}\right)\right] \end{aligned} \tag{32}$$

Cubic TSF is the last involved method, described as third degree polynomial functions for both increasing and decreasing slopes, as explained in Eq. (35).

$$f_{inc}(\theta) = \frac{3T_{ref}}{\theta_{ov}^2}(\theta - \theta_{on})^2 - \frac{2T_{ref}}{\theta_{ov}^3}(\theta - \theta_{on})^3$$
$$f_{dec}(\theta) = T_{ref} - f_{inc}(\theta - \theta_{off} + \theta_{on}) \tag{33}$$

Plotting in **Figure 5**, all the four functions superimposed for comparison some important remarks can be underlined.

As depicted in **Figure 5a**, the TSF for the cubic and sinusoidal evolutions nearly overlap and for a better comparison in **Figure 5b**, a zoomed plot of the rising slopes of the functions is depicted.

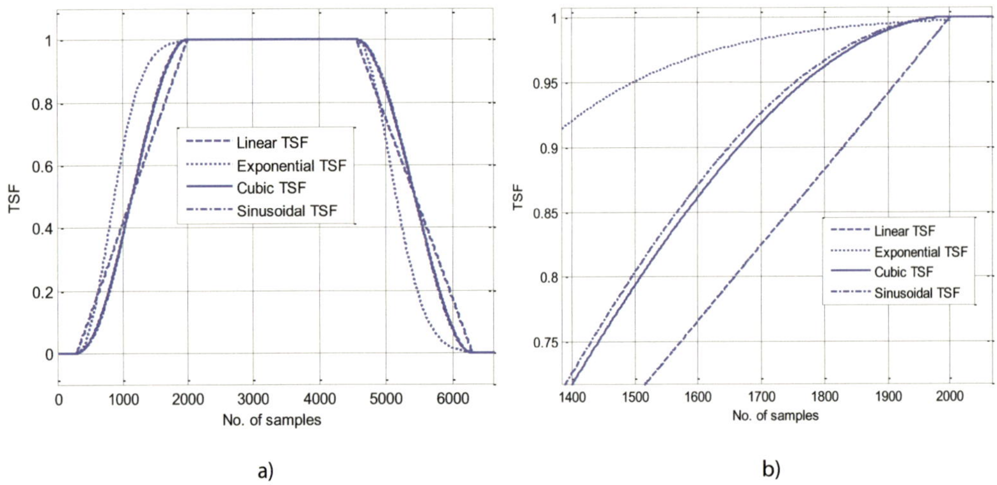

Figure 5. The variation of the TSFs used in the SRM controller.

The exponential one has the largest deviation from the linear one, the latter being considered for reference. Another important remark to be mentioned is that the machine's geometry or its parameters do not have an important significance over the effectiveness of the TSFs. Main modifications regard changing the stator to rotor pole ratio. However, it is possible to invoke optimization regarding the losses and the torque variation [17], together with the on and off switching angles.

3.3. Testing the control strategies for torque linearization

As already stated, and depicted in **Figure 2**, the torque ripples of the SRM with 'natural' hysteresis control strategy are too high to be used in EV propulsion systems [18]. However, applying the control procedures detailed in Section 3.2, this drawback can be compensated to reach a torque characteristic close to a linear one, comparable with the one of a permanent magnet synchronous or induction machine. To test the control strategies, a Matlab/Simulink model was created for the SRM based on a hybrid model based on equations and on data both fetched from the FEA model.

In **Figure 6a** and **b**, the simulation results for the DITC are depicted for an imposed speed of 1000 rpm testing at 1 and 3 Nm. The phase currents and the phase torques provide information of the contribution of each of them to the total electromechanical torque development. The latter, as seen, reaches a quite linear characteristic, which is maintained inside the hysteresis bands. One important mention that needs to be expressed is that for such simulation or real control, the computation step time must be imposed to values that allow at least 5–10 steps inside the hysteresis band. Another important issue is that the width of the bands automatically increases the switching frequency of the transistors. Hence, a compromise between the latter and the type of power switch used must be considered when sizing the bands.

Plotting the results for the linear TSF was accomplished in **Figure 6c** and **d** for the same conditions as for the DITC. There are similarities between them; however, the results for the linear TSF are smoother than those obtained with DITC. The variation of the phase torque during phase switching is quite linear, as expected for this method. Both for the DITC and the linear TSF tests were performed at low and rated torque to prove the operational skills in extreme conditions. As **Figure 6** shows that the expectations are reached in all the cases. Tests were performed ranging the rotor speed from low to rated one and still the controller responded well, linearizing the torque as expected.

The other three TSF strategies, the exponential, cubic and sinusoidal ones were tested for 1000 rpm at 3 Nm and are depicted in **Figure 7**. As seen, these too can linearize the shape of the instantaneous torque.

However, the lowest ripple is yelled by the exponential TSF while for the cubic and sinusoidal, during phase commutation there are some spikes that are visible in the plots. However, as

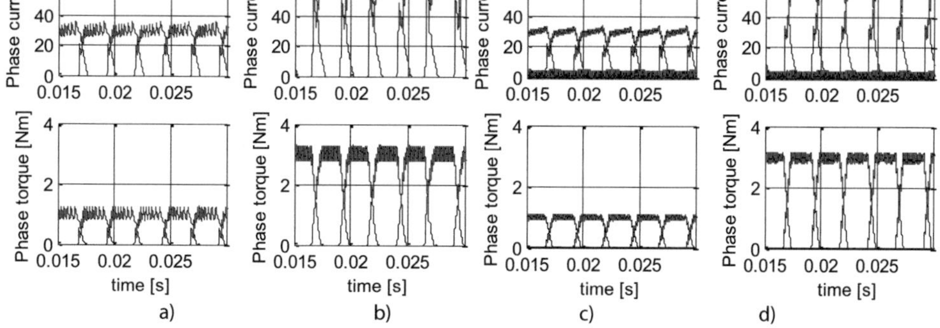

Figure 6. Simulated results for: (a) DITC @1 Nm, (b) DITC @ 3 Nm, (c) linear TSF @1 Nm and (d) linear TSF @3 Nm.

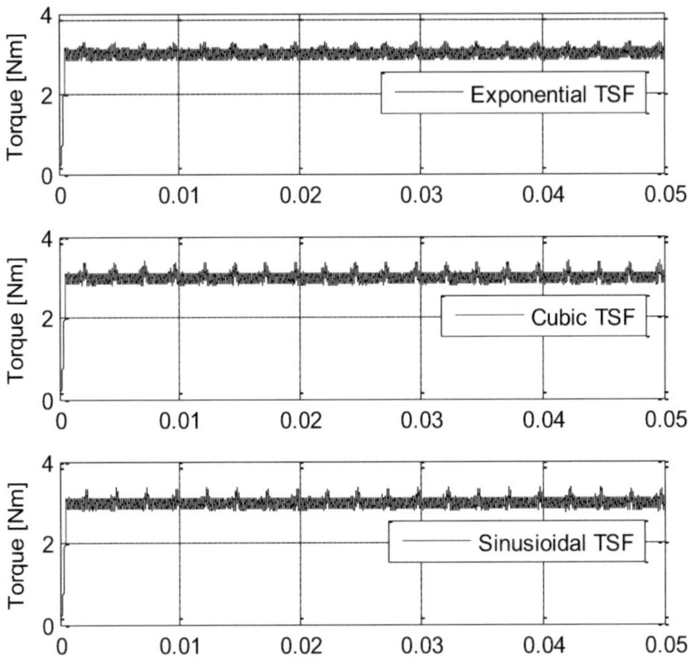

Figure 7. Simulated results for exponential, cubic and sinusoidal TSF @3 Nm.

global conclusion, the DITC and all the four TSFs can move the SRM from the point of high torque ripples to the point where these are neutralized by obtaining a linear characteristic. By this, the SRM becomes a candidate with serious advantages for the field of electric vehicle propulsion systems, combining low costs, high efficiency, fault tolerance and linear torque, over the entire range of speed and torque values.

4. Custom made SRM electronic converter architecture

In SRM-drive application, the 48 V input voltage seems to present some advantages in comparison with the 24 V systems. For application where the 48 V is not available, as for the designed SRM, this section is presenting a possible solution regarding the electronics that can enhance the performances of SRM in 24 V systems. One way to boost the input voltage by means of a front-end converter is increasing the voltage up to 48 V. For this, numerous electronic circuits can be implemented [19]. Usually, these converters are using an inductance that can increase the size of the converter, but also its price.

In this section, based on a derived C-dump topology [20], a low-cost SRM electronic circuit is presented which can add a boost voltage to a regular asymmetrical converter, increasing the overall performances of the drive system. The present circuit is fed from a 24-V power supply, as designed, and can perform in some situations like the classic SRM converter with 48 V input voltage. This is achieved, as depicted in **Figure 8**, by adding the transistor Q and the capacitor C to a regular asymmetrical SRM drive. Moreover, the input diode D is needed for proper control of the power flow. This diode is adding some losses to the circuit, but the auxiliary feature obtained by adding protection in revered polarizations is an important characteristic for automotive electronics. The auxiliary circuit is boosting the voltage across the capacitor C, by recovering the energy from the motor during de-fluxing. The voltage across C can be higher than the input voltage and this energy can be re-used to drive the motor when needed. Thus, with proper control, the motor can be fed in this situation with a higher voltage then the input voltage. If the voltage across C is regulated to a 48 V, the drive can act in some working modes like a 48-V drive system.

The energy recovered from the motor is not high enough to fed the motor all the time, but if we consider that the high voltage is strongly needed only in some particularly situations like the beginning of the phase energizing, the stored energy in the capacitor C could be sufficient.

The circuit is presented in **Figure 8**, while **Figure 9** highlights the main working modes for one phase. In the first two figures (**Figure 9a** and **b**), the phase fluxing is presented, while **Figure 9c** and **d** highlights de-fluxing possible working modes.

In the followings, two comparison situations will be presented to highlight the advantages obtained by using the presented topology in correlation with the regular asymmetrical SRM converter, applied on a 48-V SRM. In **Figure 10** and **11**, the instantaneous toque, the phase current and the phase voltage are presented at 800 and 1000 rpm, for the analyzed converter topologies.

This circuit is working with different voltage levels, thus the control of the T1 switch should be correlated with the working speed. At low speed, the energy needed to obtained the 48 V across the C is easily obtained, thus the excess energy is used during normal phase operation, or correlated with special control algorithms that can be used for smoothing the torque ripple at low speeds. At high speeds, the energy recovered in the capacitor C is just high enough for full or partial first energising of the phase coil.

If the energy recovered in the capacitor C is managed in the right way, the drive converter can act almost like a 48-V converter. This is boosting the performance of the drive system, by

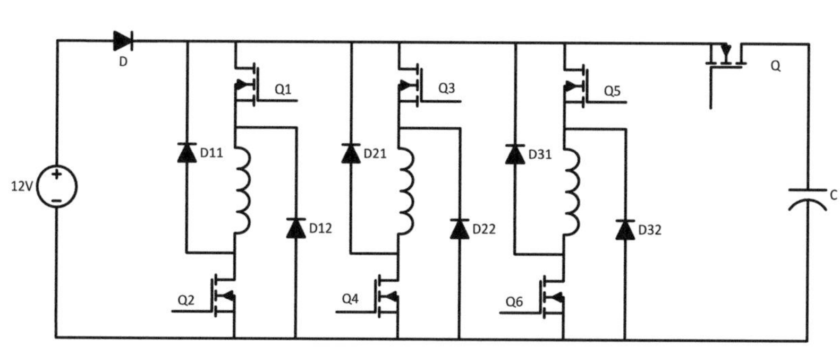

Figure 8. Proposed SRM boost converter.

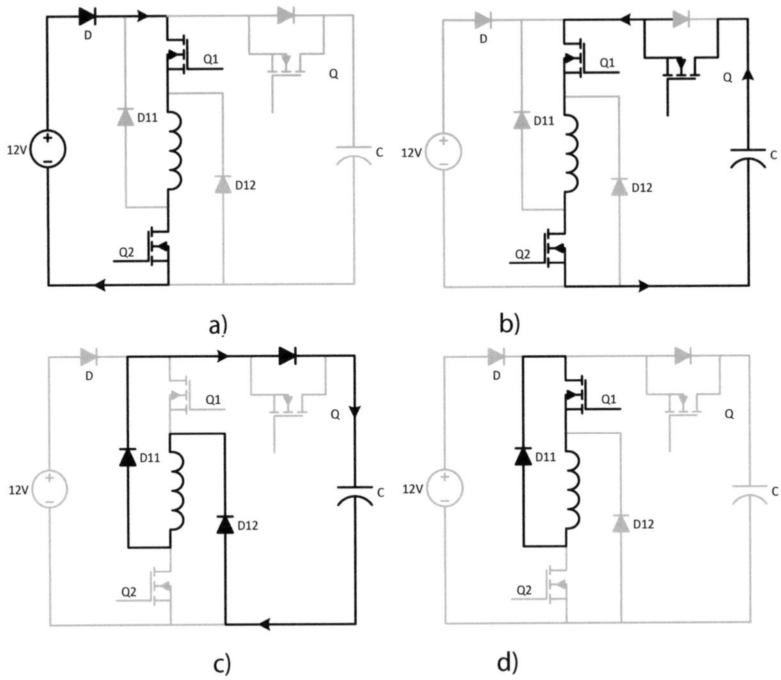

Figure 9. Operation points for the investigated converter topology; one phase is considered; (a and b) phase energising; (c and d) phase de-fluxing.

adding flexibility regarding the torque control, extend the working range and is expected to add 3–5% on the system efficiency. Moreover, the presented converter topology can be used to enhance the performances also in 12 V systems.

Comparing the results in **Figure 10** and **11**, a first remark regards the slope of the energized coil. It can be observed that due to boosted voltage, the slope of the current is faster than in the

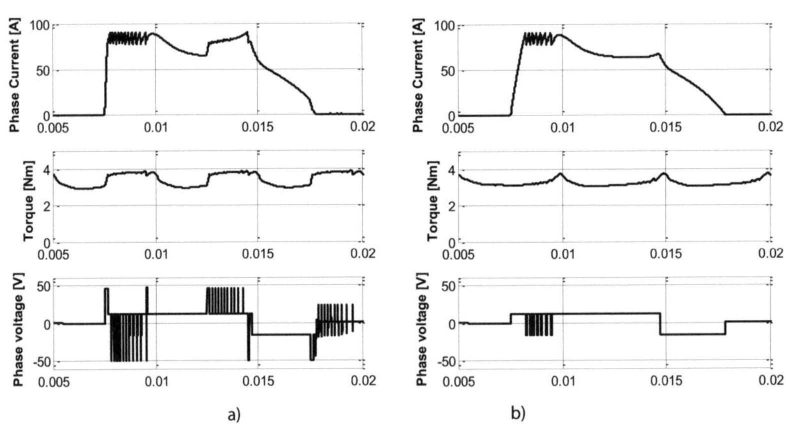

Figure 10. Waveforms for the proposed converter (a) and regular asymmetrical SRM converter (b) at 1000 rpm.

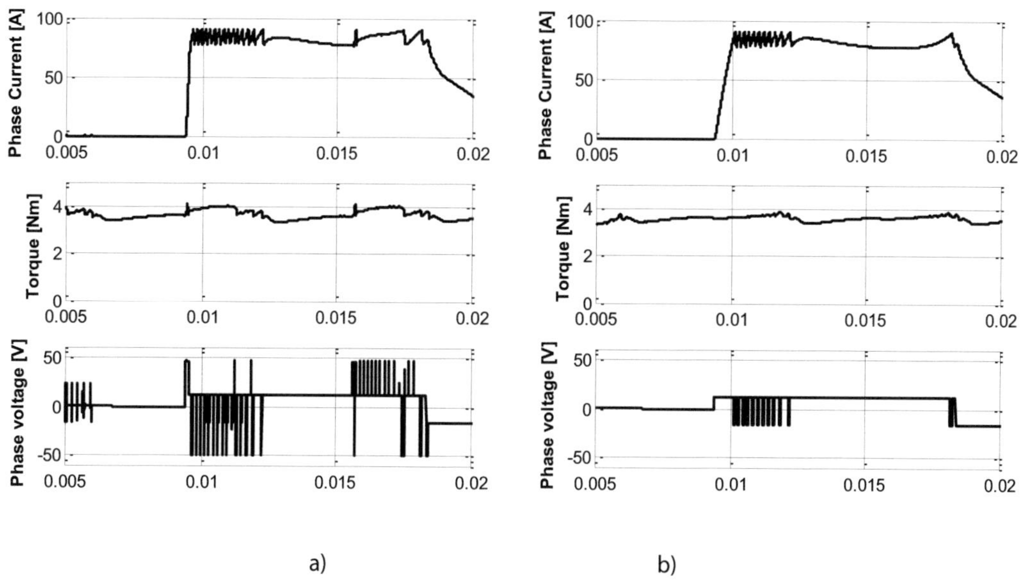

Figure 11. Waveforms for the proposed converter (a) and regular asymmetrical SRM converter (b) at 800 rpm.

case of non-boosted one. On the other hand, if comparing the mean torque, developed by the machine for all cases, at 1000 rpm this value reaches 3.36 Nm for the boosted case, instead of 3.18 Nm for the regular converter. For the second case of analysis, the mean torque in the case of the modified converter yells for 3.65 Nm while for the regular one reaches only 3.55 Nm.

Implying such changes, one can express that with low cost modifications of a classical asymmetrical converter structure, higher performances can be obtained. It is true that the gain is not extremely high, but considering that the subject is dedicated to electric propulsion systems, each step forward in increasing the autonomy of the vehicle is an addition to the actual progress of science in this field of high interest nowadays.

5. Conclusions

The goal of the chapter is to offer a solution as a complete development tool for the reader to be able to design an SRM, design an increased performance power converter for it and create smart control strategies to reach linear torque characteristics as requirement for electric propulsions. Combining analytical models of the design with finite element analysis-based validation can reach in a properly design machine. Usually, a backtracking concept is engaged, performing changes at the level of design and observe performance modifications during FEA simulations. Once the machine fits in the designer's expectations, adding an electronic power converter and torque smoothening strategies becomes the second and third steps in the design of the system.

The issues detailed in the chapter points out that an SRM can be used easily in the field of electric propulsion. Placing it in the list of serious candidates for the automotive industry, adds to the actual status of research focused on PMSMs or AC machines, with a low cost, robust and simple solution. Comparing the SRM with the above-mentioned machines, it can be stated that all need precise rotor position measurement, performed by a resolver, all need power electronics and a main electronic digital controller. However, the price of electronics nowadays decreased a lot due to fast advance. Hence, the battle on financial level is now dictated by the architecture of the machine, the materials used and the complexity of its geometry. Regarding the magnetic cores, the windings, the use of permanent magnets, etc., a comparative analysis of the ac machines with the SRM points strongly for higher costs and more complex manufacturing process.

Worldwide, there are already several companies that invest in development of SR machines for propulsion systems, both for light and heavy electrical vehicles.

Author details

Mircea Ruba* and Petre Dorel Teodosescu

*Address all correspondence to: mircea.ruba@emd.utcluj.ro

Technical University of Cluj Napoca, Cluj, Romania

References

[1] Vrazic M, Vuljaj D, Pavasovic A, Paukovic H. Study of a vehicle conversion from internal combustion engine to electric drive. In: IEEE International Energy Conference, Croatia; 13-16 May 2014. pp. 1544–1548. ISBN: 978-1-4799-2449-3

[2] Ruba M, Viorel IA, Szabó L. Modular stator switched reluctance motor for fault tolerant drive systems. IET Electric Power Applications. 2013;**7**(3):159–169. ISSN: 1751-8660

[3] Raminosoa T, Blunier B, Fodorean D, Miraoui A. Design and optimisation of a switched reluctance motor driving a compressor for a PEM Fuel cell system for automotive applications. IEEE Transactions on Industrial Electronics. 2010;vol.9, pp:2988–2997. ISSN 0278-0046

[4] Chindriş V, Ruba M, Fodorean D. Design and testing a low-voltage high-current drive for SRMs used in light electric vehicles. In: Power Electronics and Motion Control Conference and Exposition (PEMC); 16th International; 21-24 September 2014. 2014. pp. 137–142. ISSN: 978-1-4799-2060-0

[5] Radun A. Design considerations for the switched reluctance motor. Proceedings of IEEE Transactions on Industrial Applications. 1995;**31**(5):1079–1087. ISSN: 0093-9994, DOI: 10.1109/28.464522

[6] Krishnan R. Switched Reluctance Motor Drives – Modeling, Simulation, Analysis, Design, and Applications. Industrial Electronics Series, publisher: CRC Press; 2001

[7] Radun A. Analytically computing the flux linked by a switched reluctance motor phase when the stator and rotor overlap. Proceedings of IEEE Transactions on Magnetics. 2000;**36**(4):1996–2003, ISSN: 0018-9464

[8] Huang S, Luo J, Leonardi F, Lipo T. A general approach to sizing and power density equations for comparison of electrical machines. IEEE Transactions on Industry Applications. 1998;**34**(1):92–97. ISSN: 0197-2618

[9] Szabó L, Ruba M. Using Co simulations in fault tolerant Machine's study. In: Proceedings of the 23rd European Conference on Modelling and Simulation (ECMS '2009); Madrid (Spain); 2009. pp. 756–762. ISBN: 978-0-9553018-8-9

[10] Raulin V, Radun A, Husain I. Modelling of losses in switched reluctance machines. Proceedings of IEEE Transactions on Industrial Applications. 2004;**40**(6):1560–1569. ISSN: 0093-9994

[11] Petrus V, Pop AC, Martis CS, Iancu V, Gyselinck J. Direct instantaneous torque control of SRMs versus current profiling—Comparison regarding torque ripple and copper losses. In: 13th International Conference on Optimization of Electrical and Electronic Equipment (OPTIM); 24-26 May 2012. pp. 366–372. ISSN: 1842-0133

[12] Inderka RB, De Doncker RW. DITC-Direct instantaneous torque control of switched reluctance drives. In: 37th Annual Meeting IEEE IAS Conference. 2002;**39**(4):1046–4051. ISSN: 0093-9994

[13] Ruba M, Fodorean D. Development of a complete motor-drive solution for light EV based on a SRM. In: 2016 International Conference and Exposition on Electrical and Power Engineering (EPE), Iasi, Romania. pp. 197–204, ISBN: 978-1-5090-6129-7

[14] Fuengwarodsakul NH, De Doncker RW. Instantaneous torque controller for switched reluctance vehicle propulsion drives. In: 20th Electric Vehicle Symposium; November 15-19, 2003 Long Beach, California

[15] Sahoo NC, Xu JX, Panda SK. Low torque ripple control of switched reluctance motors using iterative learning. IEEE Transactions Energy Conversion. 2001;**16**(4):318–326. ISSN: 0885-8969

[16] Lee DH, Liang J, Lee ZG, Ahn JW. A simple nonlinear logical torque sharing function for low-torque ripple SR drive. IEEE Transactions on Industrial Electronics. 2009;**56**(8):3021–3028. ISSN: 0278-0046

[17] Mademlis C, Kioskeridis I. Performance optimization in switched reluctance motor drives with online commutation angle control. IEEE Transactions on Energy Conversion. 2003;**18**(3):448–457

[18] Ruba M, Fodorean D. Motor-drive solution for light electric vehicles based on a switched reluctance machine. In: 2016 IEEE International Conference on Automation, Quality and Testing, Robotics, AQTR; 19th-21st May 2016; Cluj Napoca, Romania. ISBN: 978-1-4673-8691-3

[19] Tomaszuk A, Krupa A. High efficiency high step-up DC/DC Converters a Review. Bulletin of the Polish Academy of Sciences Technical Sciences. 2011;**59**(4):475–483. ISSN: 2300-1917

[20] Lee TW, Yoon YH, Yuen-Chung Kim, Lee BK, Won CY. Control of c-dump converters fed switched reluctance motor on an automotive application. Electric Power Systems Research. 2007;**77**(7):804–812. ISSN 0378-7796

Chapter 5

Current-Controlled SRM Fed by Three-Phase Boost PFC

Erdal Şehirli and Meral Altınay

Additional information is available at the end of the chapter

http://dx.doi.org/10.5772/intechopen.69150

Abstract

In this chapter, firstly, converter types of switched reluctance motor (SRM) are described. Current control structure of SRM, which has six stator and four rotor poles, over an asymmetric bridge converter, is also explained. While feeding SRM by an AC grid, grid voltages have to be converted to DC voltage; to realize this conversion, in order to obtain high power factor and sinusoidal grid current, power factor correction (PFC) circuits must be used. In this study, an asymmetric bridge converter of SRM is fed by three-phase PFC boost converter that consists of uncontrolled diode rectifier and DC-DC boost converter with high frequency operation. PFC boost converter is controlled by nonlinear control algorithm. By means of the simulations that are conducted by MATLAB/Simulink, grid voltage and current, current harmonics of each phase, three-phase currents of phases, flux, and current of SRM are presented. Simulation results show that proposed SRM that is fed by three-phase PFC boost converter system gives the desired performance, for both grid and SRM side.

Keywords: converter, PFC, current control, SRM

1. Introduction

Recently, using SRM has taken much attention because of improving semiconductor and microprocessor technology. Furthermore, SRM has some advantages that are having concentric stator winding and salient stator, rotor poles without any winding, and magnets. Besides, with respect to induction motor drivers, SRM driver needs lesser switch. Another advantage of SRM is to have constant torque characteristic. On the other hand, one of the important disadvantages of SRM is torque ripple. But torque ripple can be improved with converter types.

SRM is not operated by using only DC source. To drive SRM, converter topologies such as asymmetric bridge, r dump, c dump are required. Furthermore, SRM converter needs DC power to operate. This DC power can easily be obtained by using an AC grid that can be single- or three-phase over rectifiers. However, if the AC grid is converted to DC by uncontrolled rectifiers, grid current has harmonics and the power factor of AC grid decreases. To overcome these problems, improved power quality converter that contains pulse width modulated (PWM) rectifiers and power factor correction (PFC) circuits should be used.

In literature, there are so many publications about SRM that are related with construction, operation, control, its converters, and obtaining a high power factor while feeding SRM. Some of the publications are summarized as follows.

In Refs. [1, 2], fundamental information about control, design, operation, and converter types of SRM is given. Converter and driver types for SRM are also expressed and compared [3, 4]. Studies on control of SRM that are about speed, torque, and current control are implemented [5–9]. Control topologies, such as adaptive, fuzzy, sliding, and nonlinear are realized for SRM [10–12].

Some new applications of SRM that are battery chargers, water pumping systems, electrical vehicles, and solar photo voltaic (PV)-powered drives are explained [13–16].

Krishnan and Lee [17] implement C dump converter with four-phase SRM, and Consoli et al. [18] is for three-phase SRM that is fed by single-phase PFC boost converter, and they provide higher power factor. Single-phase half-controlled PWM boost converter-based PFC converter is used for feeding four phases of SRM that is driven by an asymmetric bridge converter as in Ref. [19] in order to obtain high power factor, and it has DC-DC converter between PFC converter and asymmetric bridge converter. Another application that obtains a high power factor, with asymmetric bridge converter, is realized for four phases of SRM by using single-phase PFC boost converter [20]. Reinert and Schroder [21] emphasize the need of PFC circuit for SRM by pointing out single-phase boost PFC and asymmetric bridge converter. Single-phase PFC boost converter is included with hysteresis current-controlled SRM drive [22], and the same study with fuzzy tuned proportional-integral-derivative (PID) controller for PFC converter is made in Ref. [23]. Rajesh and Singh [24] present SRM with midpoint converter that is fed by Vienna rectifier for having higher power factor. SRM is also driven by midpoint converter and fed by single-phase three-level rectifier as a PFC circuit [25]. To have reversible power flow and regenerative specification [26], two different PFC converters that are active power filters (APC) and three-phase voltage sourced converters (VSC) are implemented. Power quality is improved by using Zeta converter [27] and canonical switching cell converter [28] for midpoint converter-driven SRM.

In this chapter, converter types of SRM are described. Current control structure of SRM which has six stator and four rotor poles, over asymmetric bridge converter, is also explained. Furthermore, in this study, asymmetric bridge converter of SRM is fed by three-phase PFC boost converter that consists of uncontrolled diode rectifier and DC-DC boost converter with high frequency operation. PFC boost converter is controlled by nonlinear control algorithm. By means of the simulations that are conducted by MATLAB/Simulink, grid voltage and current, current harmonics of each phase, three-phase currents of phases, flux, and current of SRM are presented. Simulation results show that proposed SRM that is fed by three-phase PFC boost converter system gives the desired performance, both grid and SRM side.

2. Converter types for SRM

SRM is not operated without using converter topologies. In literature, SRM converters are classified as shown in **Figure 1** [1]. This classification is based on the switch number in converter topology.

Some of the converter topology from each classification, that are two-stage power converters, asymmetric bridges, R dump converter, and C dump converter, is expressed briefly as follows.

2.1. Two-stage converter

In a two-stage converter, power conversion is realized as bidirectional that is transferring energy from SRM to grid and from grid to SRM. The block diagram of two-stage topology is shown in **Figure 2** [1]. In **Figure 2**, it has been shown that when the SRM is used as a generator, prime mover is attached to SRM and inverter is used to obtain three-phase voltage for grid integration. However, when the SRM is used as a motor, rectifier is used to feed SRM from the three-phase grid and load is replaced with prime mover.

2.2. 1.5 switch/phase converter

1.5 q converter that is shown in **Figure 3** [1] is just suitable for SRM that has even number of phases. Energizing for each phase, 1.5 switch and diodes are required. The operation of 1.5 switch converter is described as follows. To energize phase a, S_1 and S_5 is turned on, while S_3 and S_6 is turned off. At that time, phase c current freewheels over D_5. After that, for energizing phase b, S_2 and S_6 have to be turned on when S_5 has to be turned off and current of phase a flows from S_1 to D_5. Then, phase c should be energized by turning on S_3 and S_5 while turning off S_6 and S_1. Furthermore, phase b current flows from S_2 to D_6 and phase a current freewheels

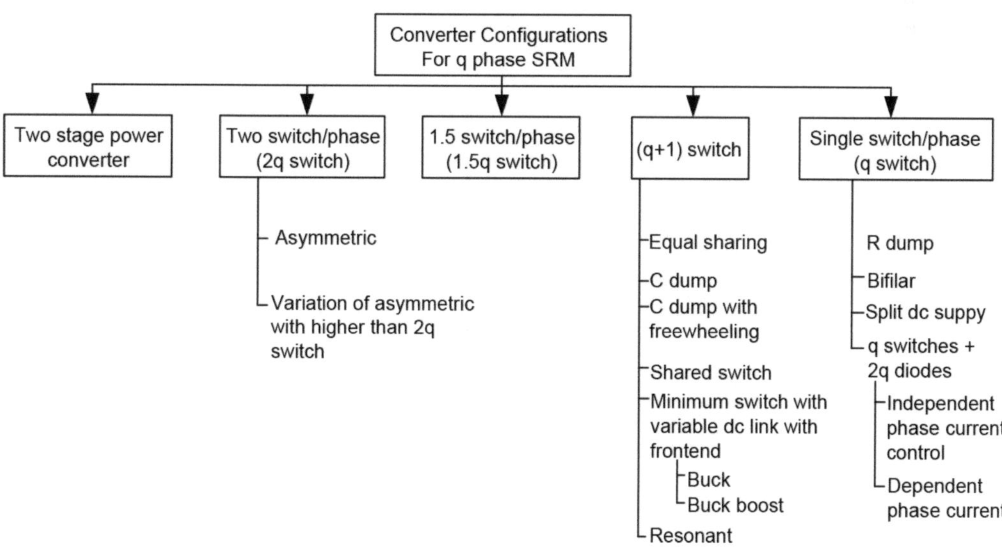

Figure 1. The converter classification of SRM [1].

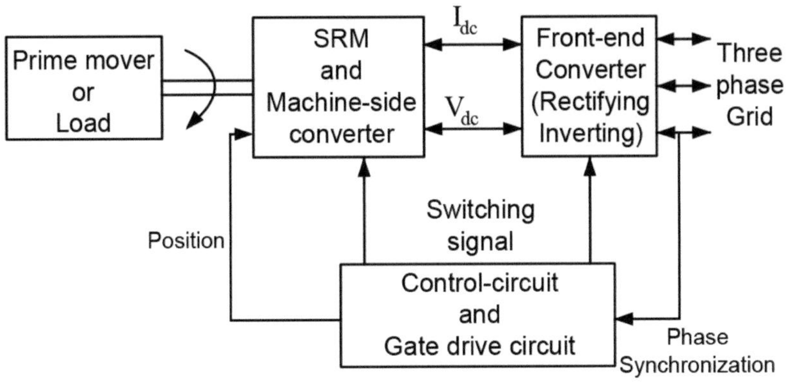

Figure 2. A block diagram of a two-stage converter.

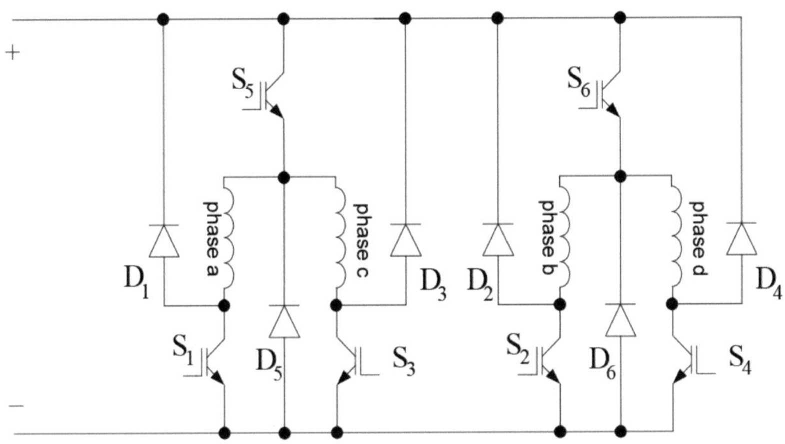

Figure 3. 1.5 switch/phase converter.

over D_1. At last, to energize phase d, S_6 and S_4 should be turned on after turning off S_5 and S_2. Besides, phase c current flows from S_3 to D_5, while phase b current freewheels over D_2.

2.3. R dump converter

R dump converter is shown in **Figure 4**. Operation of R dump converter can be summarized as follows. After each power switch turns off, current passes through the diodes and charge C; then, current flows through R and resets the energy in phase windings.

Due to the use of R for extinguishing energy of phase windings, energy is dissipated and overall efficiency reduces.

2.4. C dump converter

Figure 5 shows a C dump converter. In C dump converter, energy recovery process is being done by S_4, D_4, L, and C. Operation is briefly as follows: firstly the energy of phase windings is

http://dx.doi.org/10.5772/intechopen.69150

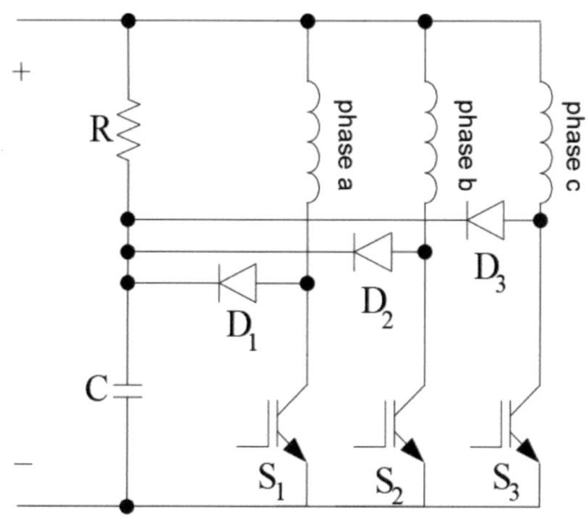

Figure 4. R dump converter of SRM.

stored on C and then by switching on S_4, stored energy on C is sent to DC source or phase windings over L and D [1, 29].

2.5. Asymmetric bridge converter

In asymmetric bridge converter that is shown in **Figure 6**, for each phase of SRM, there are two power switches and two diodes.

The sample connection of switches and diodes to one phase of SRM is shown in **Figure 7**. It is also seen from the same figure that SRM has six stator and four rotor poles. It is the same structure as the study that is carried out in this chapter.

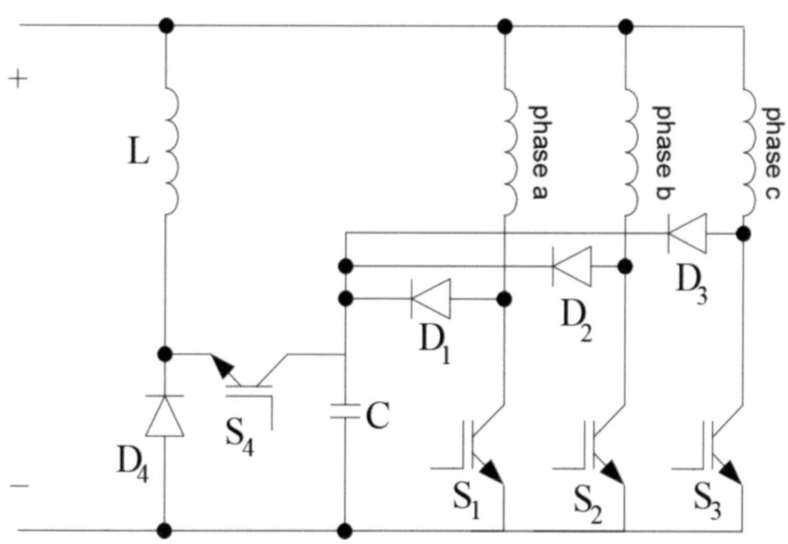

Figure 5. C dump converter of SRM.

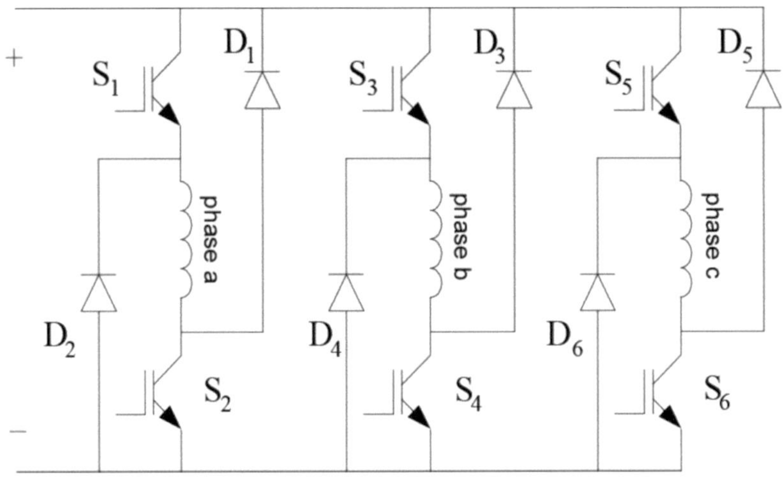

Figure 6. Asymmetric bridge converter of SRM.

Basically, the definition of **Figure 7** and the operation of asymmetric bridges are as follows: windings are placed mutually on stator poles and they are connected in series to compose N and S poles. After energizing phase one winding with S_1 and S_2 switches, current flows from DC power supply to phase winding. When switches turn off, current flows through D_1 and D_2 diodes and negative voltage is applied to phase winding that causes decreasing of winding current. So, energy turns back to power supply [30].

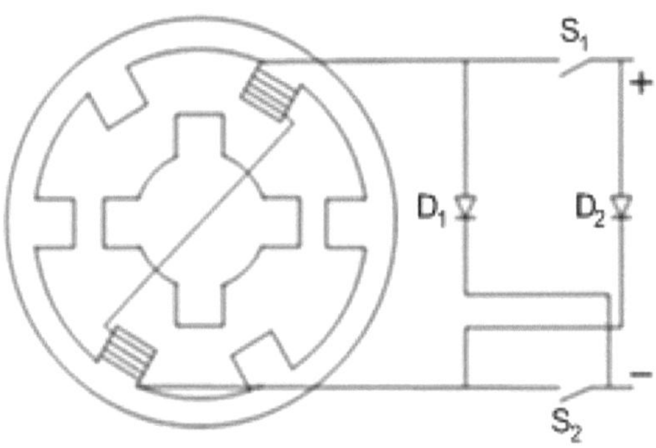

Figure 7. Connection of an asymmetric bridge to SRM.

3. Current control of SRM

In literature, in order to control speed of SRM, current control of SRM has to be done as a first stage that is seen in generalized speed control block diagram in **Figure 8**.

Furthermore, current control of SRM can be realized by hysteresis and PWM methods [30]. In this chapter, PWM-based control is chosen as a current controller. In **Figure 9**, current control block diagram is shown.

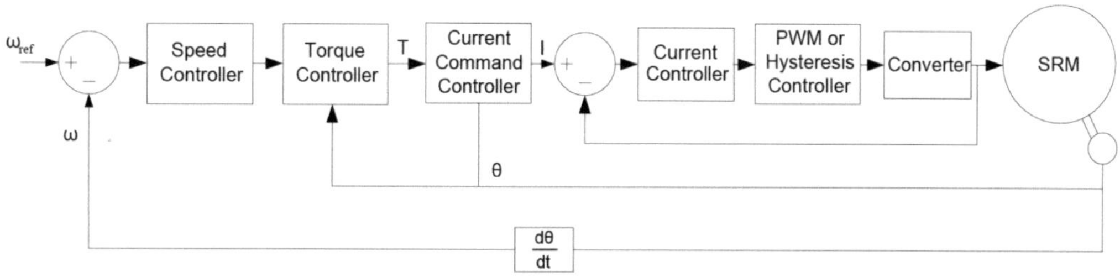

Figure 8. A generalized speed control block diagram of SRM.

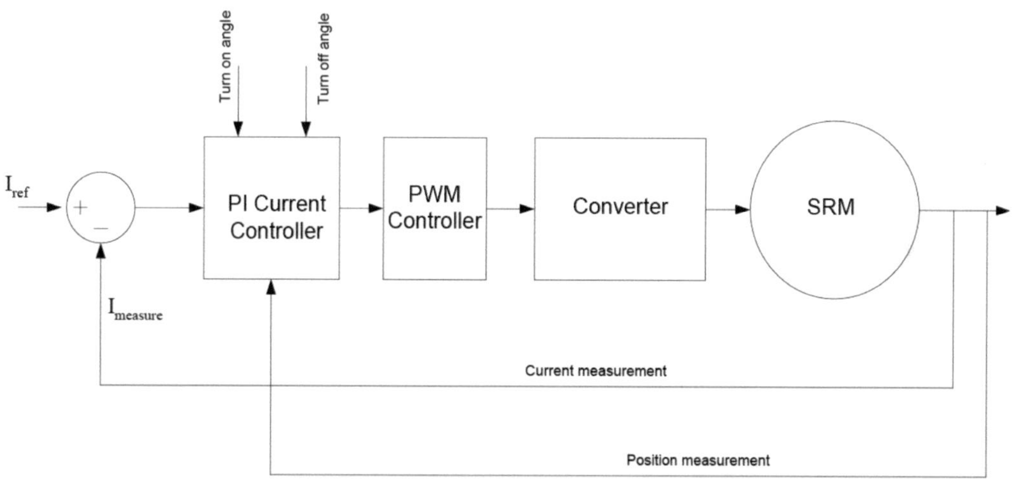

Figure 9. Current control block diagram of SRM.

In **Figure 9**, it is seen that, winding currents of SRM are measured and compared with the reference current. After current comparison, current differences are sent to PI current controller block. PI current controller block produces appropriate control signals with respect to turn-on and turn-off angle of SRM's rotor position. Then, these signals are applied to PWM block in order to obtain gating pulses of the asymmetric bridge converter. After applying pulses to asymmetric bridge converter, SRM is driven [30].

4. Three-phase PFC boost converter

SRM can be fed after rectifying the AC grid. This rectification process can be done by using wide variety of rectifiers that are conventional rectifiers and increased power factor corrected converters. In **Figure 10**, all the converters that are in literature [31] are classified.

However, conventional rectifiers that use uncontrolled and controlled thyristors do not provide a high power factor and sinusoidal grid currents. So, with conventional topologies, AC grid is not used efficiently. Therefore, in order to obtain high power factor and sinusoidal grid

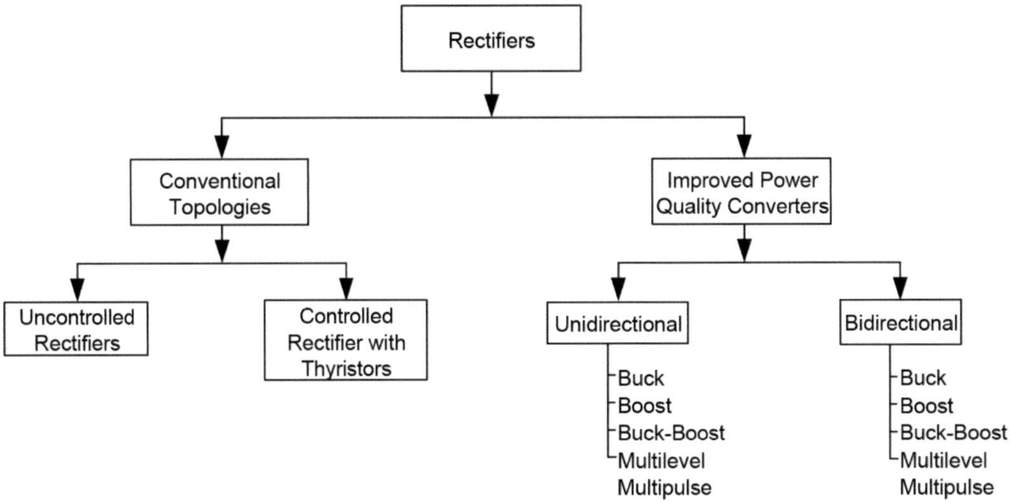

Figure 10. Classification of converters that can be used as front-end converter for SRM.

currents, increased power factor correction topologies (IPQC) are chosen. IPQC topologies can also be classified with respect to the power flow as unipolar and bipolar topologies [31].

By using the bipolar topologies, SRM can also be operated as a motor and generator, or regenerative operation of SRM is provided. But bipolar topologies include many power switches. So, it increases the cost of the converter. On the other hand, unipolar topologies include less power switch and the cost can be minimized. Also, implementation of unipolar converters is not as complex as bipolar converters.

In literature, some special properties of unipolar converters are generally called as PFC converters [32]. Furthermore, PFC converters can be realized by using any DC-DC converter, such as buck, boost, and buck-boost that are connected to the output of uncontrolled rectifier. This rectifier can be single or three phase. By operating the switch of DC-DC converter with the right control algorithm, a high power factor with sinusoidal grid current is obtained.

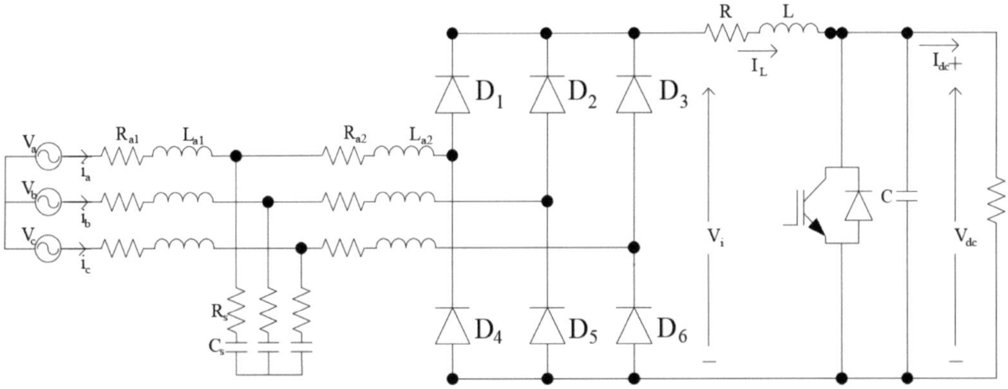

Figure 11. A three-phase PFC boost converter.

In **Figure 11**, three-phase PFC boost converter that has six diodes and one power switch is shown. It is seen in **Figure 11** that three-phase PFC boost converter is connected to the grid over inductor-capacitor-inductor (LCL) filter. In the LCL filter, series resistors to capacitors are connected in order to provide damping, and series resistor is used to show internal resistors of inductances.

A mathematical model of three-phase PFC boost converter can be written in Eq. (1) by applying Kirchhoff's current and voltage law with respect to the on and off position of the switch [32].

$$\begin{bmatrix} \dot{i}_L \\ \dot{V}_{dc} \end{bmatrix} = \begin{bmatrix} 0 & (-1+\mathrm{d})/\mathrm{L} \\ (1-\mathrm{d})/\mathrm{C} & -1/\mathrm{RC} \end{bmatrix} \begin{bmatrix} i_L \\ V_{dc} \end{bmatrix} + \begin{bmatrix} 1/\mathrm{L} \\ 0 \end{bmatrix} V_i \quad (1)$$

After applying nonlinear control method that is defined in Refs. [32–34], nonlinear controller for three-phase PFC boost converter is obtained in order to regulate DC voltage of the asymmetric bridge converter of SRM.

In **Figure 12**, nonlinear controller of three-phase PFC boost converter is shown as a block diagram.

Nonlinear controller in **Figure 12** needs to have reference current (I_{ref}), measured voltage (V_{dc}), and inductor current (I_L). Furthermore, reference current (I_{ref}) is produced by PI controller after comparing the square of reference and measured voltage [35]. Then, nonlinear control law is applied inside of nonlinear PFC controller block [30, 32]. After that, a new control variable (u) is obtained. By means of the comparison between u and a saw tooth wave form that has 100-kHz frequency in DC-PWM block, PWM pulse for power switch of PFC boost converter is acquired. After applying this PWM pulse to switch, required DC voltage is obtained.

In **Figure 13**, nonlinear controlled three-phase PFC boost converter is shown.

Nonlinear PFC controller for three-phase PFC boost converter is realized by using the Eq. (2) as in Refs. [30, 32–35].

$$\mathrm{u} = -\mathrm{K}(V_{dc} - V_{REF})\frac{C}{I_L} + 1 - \frac{V_{dc}}{RI_L} \quad (2)$$

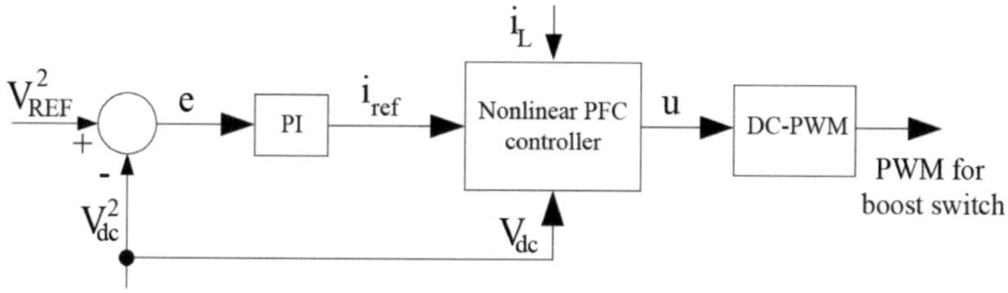

Figure 12. A block diagram of nonlinear controller of PFC boost converter.

5. Simulations

Current-controlled SRM that is fed by three-phase PFC boost converter is realized by simulation. In this simulation, Matlab/Simulink is used as in Refs. [36, 37]. A block diagram of simulation is shown in **Figure 14**.

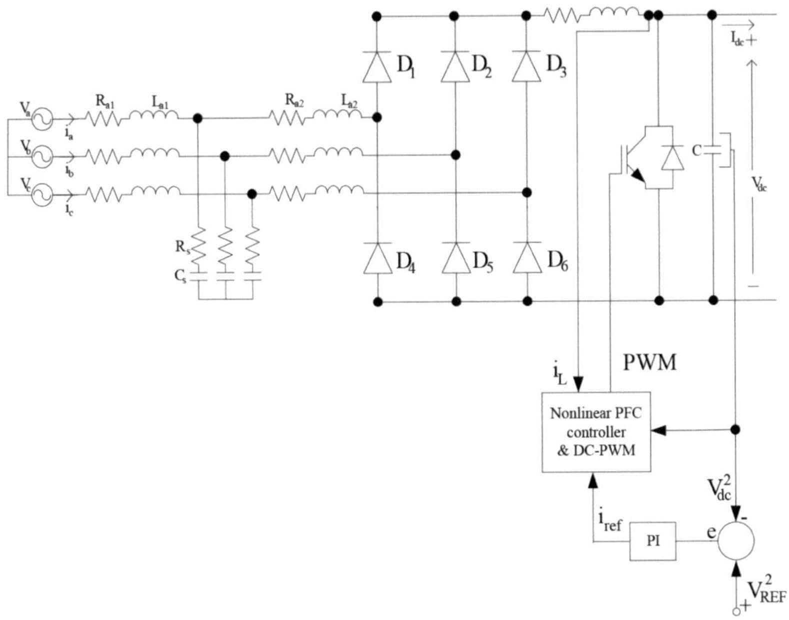

Figure 13. A block diagram of nonlinear controlled three-phase PFC boost converter.

In **Figure 14**, a three-phase AC grid is rectified by three-phase uncontrolled rectifier that is connected to grid by LCL filter. Besides, to build PFC structure, a boost converter that is operated with high switching frequency is added after the rectifier. PFC boost converter is controlled by nonlinear control structure and uses the DC-PWM technique to produce PWM pulse for power switch. Also, it is used to adjust the DC voltage of asymmetric bridge converter that is connected to SRM phase windings.

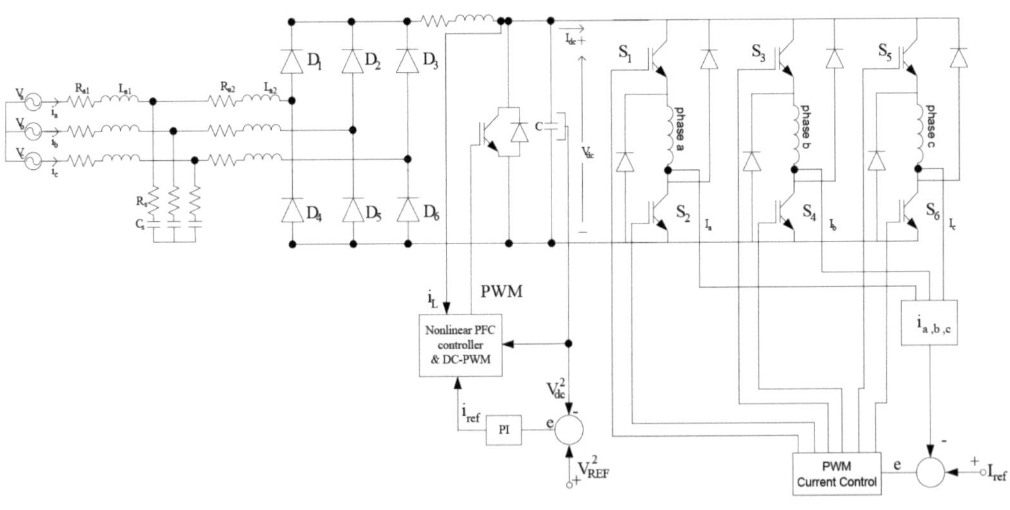

Figure 14. A block diagram of current-controlled SRM fed by three-phase PFC boost converter.

http://dx.doi.org/10.5772/intechopen.69150

In **Figure 15**, simulation circuit is shown. In this simulation, SRM current and flux are measured with reference changes. Also, three-phase current and their harmonics, single-phase voltage, and current are shown. Furthermore, SRM is loaded with 10 Nm as a load.

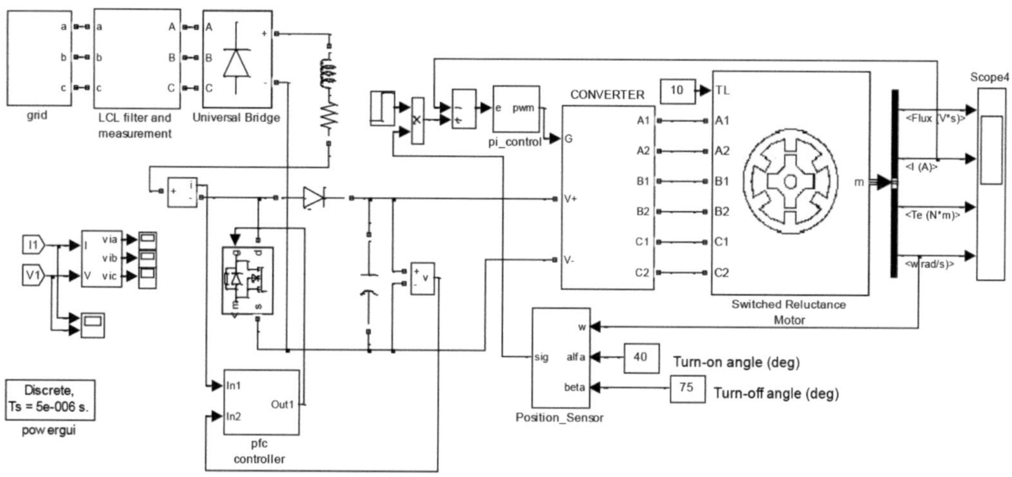

Figure 15. Simulation circuit of current-controlled SRM fed by three-phase PFC boost converter.

The parameters that are used in simulations are given in **Table 1**.

In **Figure 16**, SRM current is shown. In that figure, it is understood that SRM current reference is changed from 20 to 30 A at 0.5 s with maximum ±5 A ripple, and current is controlled as desired.

In **Figure 17**, SRM flux is shown under reference change of SRM current. After current reference increases, it is observed that flux increase as desired.

Passive components							
LCL filter						**DC link**	
R_{a1} (Ω)	L_{a1} (μH)	R_{a2} (Ω)	L_{a2} (H)	R_s (Ω)	C_a (F)	R_L (Ω)	C_{dc} (μF)
2	0.1	0.01	0.003	0.15	0.0023	1000	2200
SRM							
Poles		Stator resistance (Ω)	Inertia (kg.m)	Friction (Nms)	Current control	PI	
Stator 6	Rotor 4	0.05	0.05	0.02		K_p 5	K_i 10
Three-phase PFC boost controller							
DC link				Nonlinear controller			
K_p		K_i		K (10^9)			
1		10		1			
f (sw)	PFC boost	69 kHz		Asymmetric bridge		9 kHz	

Table 1. Parameters used in simulations.

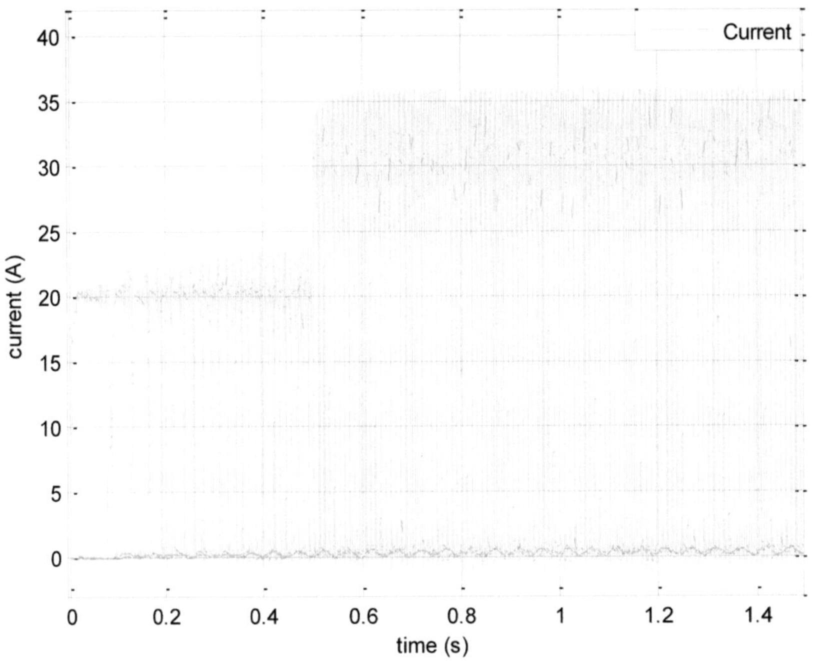

Figure 16. SRM current.

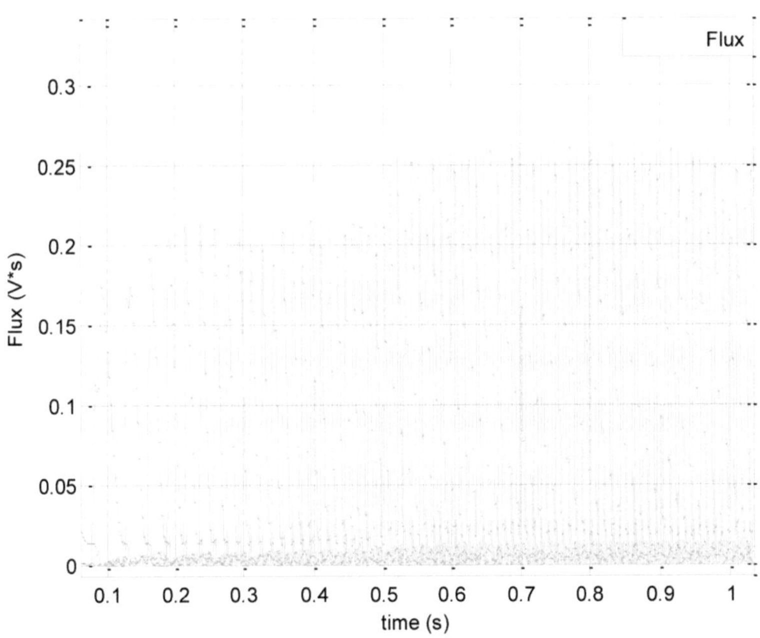

Figure 17. SRM flux.

http://dx.doi.org/10.5772/intechopen.69150

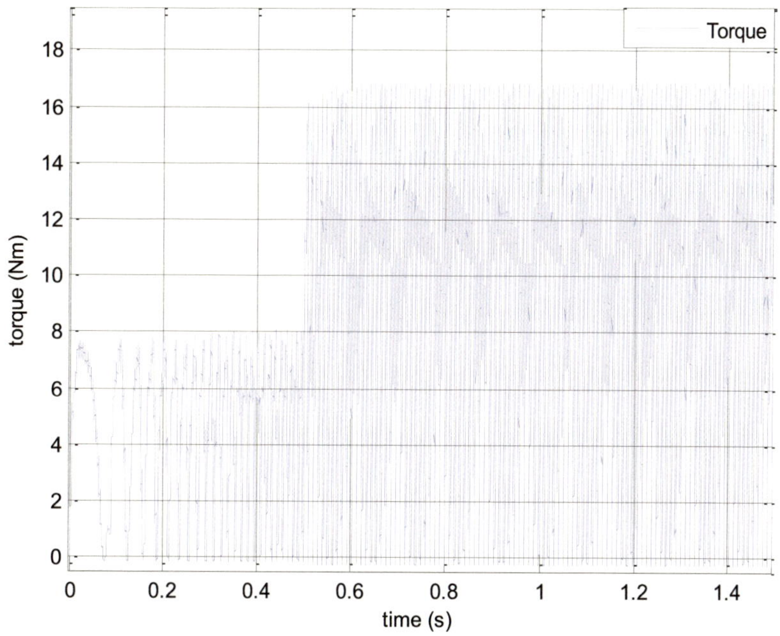

Figure 18. SRM torque.

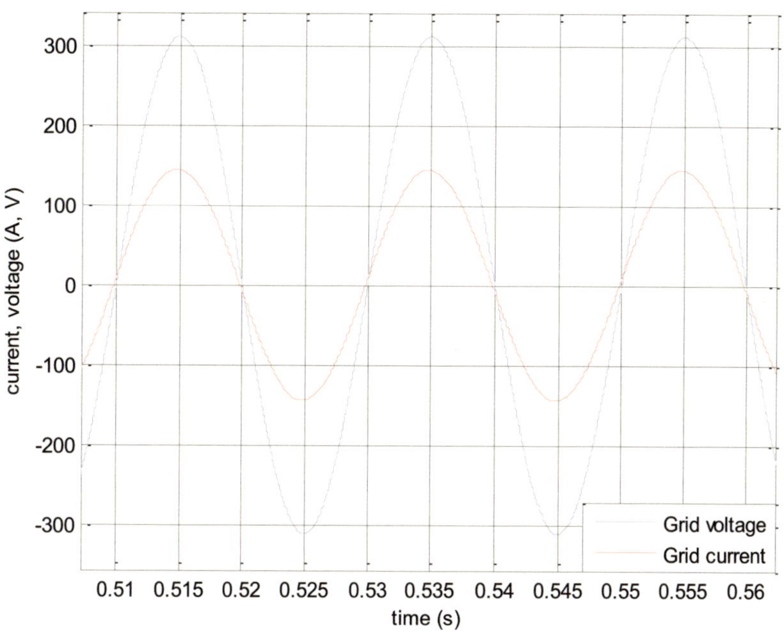

Figure 19. Single-phase current and voltage of grid.

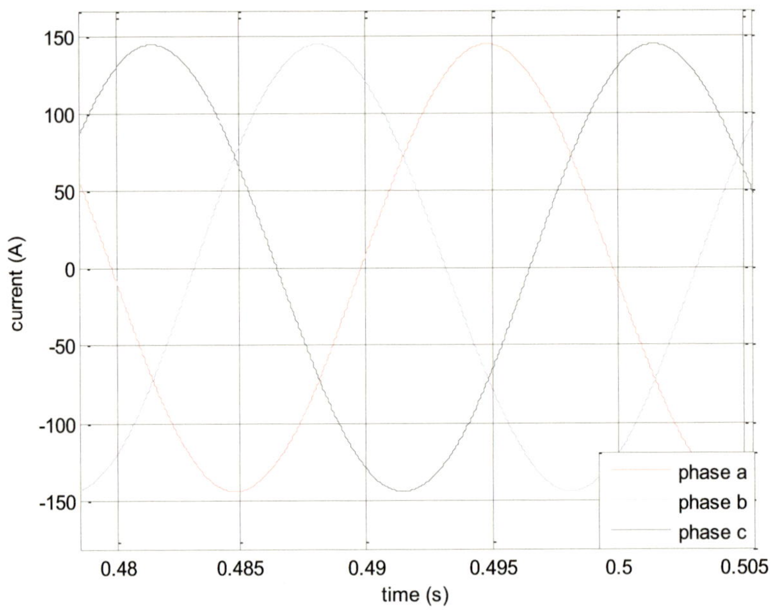

Figure 20. Three-phase currents of grid.

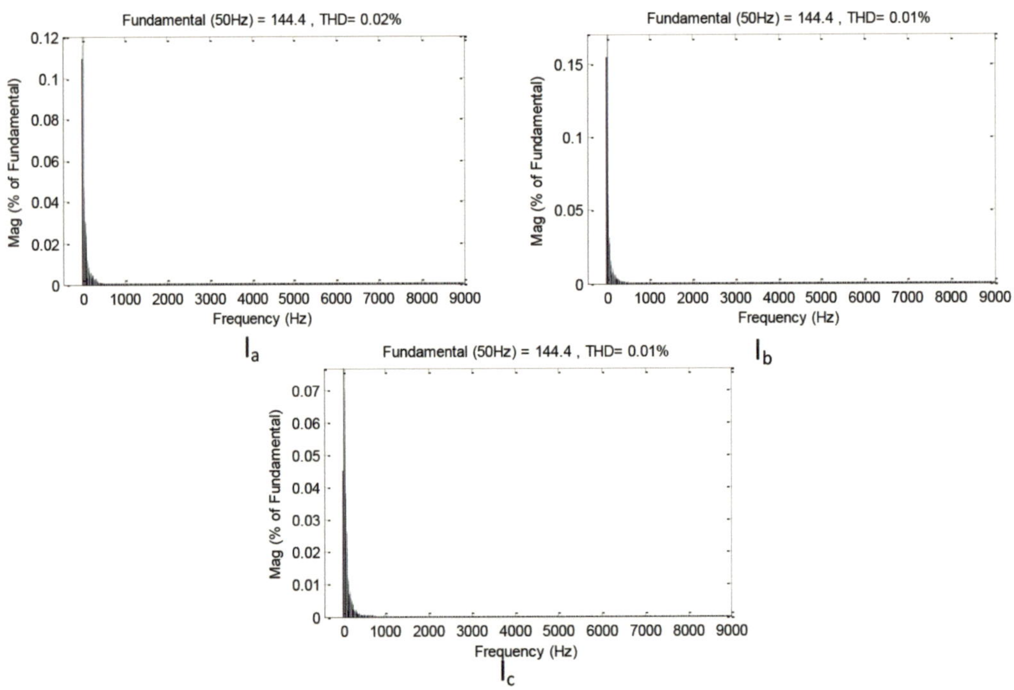

Figure 21. THDs of grid currents (Ia, Ib, Ic).

SRM torque per phase is shown in **Figure 18**. When current command is changed at 0.5 s, electromagnetic torque per phase of SRM is also changed.

Single-phase current and voltage are shown in **Figure 19**. It is observed that there is no phase shift between phase current and voltage. So, power factor is said to be unity. Therefore, there is no reactive power drawn from the grid.

In **Figure 20** three-phase currents are shown. It is seen that three-phase currents are sinusoidal and balanced.

Total harmonic distortion (THD)'s of three phase currents are shown in **Figure 21**. It is seen that THDs are 0.02, 0.01, and 0.01% for each phase, respectively. It can be understood by **Figures 19–21** that three-phase PFC boost converter operates as desired by providing DC power to SRM, lower THDs on grid current, and unity power factor.

6. Conclusion

In this chapter, firstly, SRM converter types that are commonly used such as asymmetric bridge, R dump, and C dump are expressed. Then, in order to supply DC power to SRM, rectifiers in literature are classified and three-phase PFC boost converter is chosen as a rectifier.

Furthermore, current-controlled SRM that is driven by using asymmetric bridge converter is realized while feeding SRM with three-phase PFC boost converter. Three-phase PFC boost converter is also controlled by a nonlinear method. This application is conducted over simulation which is realized with MATLAB/Simulink. By means of the simulation results, THDs of grid currents are obtained as 0.02, 0.01, and 0.01% for each phase, respectively. Also, grid voltage and grid currents are in the same phase so unity power factor is obtained. As a result, SRM current is controlled as desired as well.

Author details

Erdal Şehirli[1]* and Meral Altınay[2]

*Address all correspondence to: esehirli@kastamonu.edu.tr

1 Vocational School for Higher Education of Kastamonu University, Kastamonu, Turkey

2 Technology Faculty of Kocaeli University, Kocaeli, Turkey

References

[1] Krishnan R. Switched Reluctance Motor Drives. 1st ed. New Jersey: CRC Press; 2001

[2] Kuyumcu FE. Switched Reluctance Machine. 1st ed. Kocaeli: Lecture Notes; 2013

[3] Ha K, Lee C, Kim J, Krishnan R, Oh SG. Design and development of low cost and high efficiency variable speed drive system with switched reluctance motor. IEEE Transaction on Industrial Application. 2007;**43**(3):703–713. DOI: 10.1109/TIA.2007.895744

[4] Vukosavic S, Stefanovic VR. SRM inverter topologies: A comparative evaluation. IEEE Transaction on Industrial Application. 1991;**27**(6):1034–1047. DOI: 10.1109/28.108453

[5] Yuan G. Speed control of switched reluctance motors [thesis]. Hong Kong: The Hong Kong University of Science and Technology; 2000

[6] Bae HK. Control of switched reluctance motors considering mutual inductance [dissertation]. Virginia: Virginia Polytechnic Institute and State University; 2000

[7] Sahoo NC, Elamvazuthi I, Shaikh RA, Nor NM. A comparative study of Single Pulse Voltage and Hysteresis current control methods for switched reluctance motors. In: Energy, Automation, and Signal (ICEAS), 2011 International Conference on; 28–30 December; Bhubaneswar, Odisha, India: IEEE; 2011. p. 1–6. DOI: 10.1109/ICEAS.2011.6147145

[8] Maroon C, Garcia A, Tremps E, Somolinos JA. Torque control of switched reluctance motors. IEEE Transactions on Magnetics. 2012;**48**(4):1661–1664. DOI: 10.1109/TMAG.2011.2173169

[9] Deihimi A. Switched reluctance machine synthesis algorithms based on PWM control strategy for machine design optimization. In: 2008 11th International Conference on Optimization of Electrical and Electronic Equipment; 22–24 May; Brasov, Romania: IEEE; 2008. p. 51–56. DOI: 10.1109/OPTIM.2008.4602386

[10] Abut N, Cakir B, Inanc N, Yıldız AB, Bilgin MZ. Switched reluctance motor drive by using fuzzy controller. In: 23rd Industrial Electronic Control and Instrumentation. In: Industrial Electronics, Control and Instrumentation, 1997. IECON 97. 23rd International Conference on; 14 Nov.; New Orleans, LA, USA: IEEE; 1997. p. 348–353. DOI: 10.1109/IECON.1997.671075

[11] Amor LB, Dessaint LA, Akhrif O, Oliver G. Adaptive input output linearization of a switched reluctance motor for torque control. In: Industrial Electronics, Control, and Instrumentation, 1993. Proceedings of the IECON '93, International Conference on; 15–19 Nov.; Maui, HI, USA: IEEE; 1993. p. 2155–2160. DOI: 10.1109/IECON.1993.339410

[12] Li Y, Tang Y, Ji-bin C, Ai-hua L. Continuous sliding mode control and simulation of SRM. In: Cognitive Informatics & Cognitive Computing (ICCI*CC), 2011 10th IEEE International Conference on; 18–20 Aug.; Banff, AB, Canada: IEEE; 2011. p. 314–317. DOI: 10.1109/COGINF.2011.6016158

[13] Hu Y, Gan C, Cao W, Li C, Finney SJ Split converter-fed SRM drive for flexible charging in EV/HEV applications. IEEE Transaction on Industrial Electronics. 2015;**62**(10):6085–6095. DOI: 10.1109/TIE.2015.2426142

[14] Mishra KA, Singh B. SPV array powered zeta converter fed SRM drive for water pumping. In: 2015 IEEE Power, Communication and Information Technology Conference (PCITC); 15–17 Oct; Bhubaneswar, India. Bhubaneswar, India: IEEE; 2015. p. 476–482. DOI: 10.1109/PCITC.2015.7438213

[15] Gan C, Wu J, Yang S, Cao W, Guerroro JM. New integrated multilevel converter for switched reluctance motor drives in plug-in hybrid electric vehicles with flexible energy conversion. IEEE Transactions on Power Electronics. 2017;**32**(5):3754–3766. DOI: 10.1109/TPEL.2016.2583467

[16] Hu Y, Gan C, Cao W, Fang Y, Finney SJ. Solar PV-powered SRM drive for EVs with flexible energy control functions. IEEE Transactions on Industry Applications. 2016;**52**(4):3357-3366. DOI: 10.1109/TIA.2016.2533604

[17] Krishnan I, Lee S. Effect of power factor correction circuit on switched reluctance motor drives for appliances. In: Applied Power Electronics Conference and Exposition, 1994. APEC '94. Conference Proceedings 1994., Ninth Annual; 13-17 Feb; Orlando, FL, USA: IEEE; 1994. p. 83-89. DOI: 10.1109/APEC.1994.316415

[18] Consoli A, Testa A, Aiello N, Gennaro F, Presti ML. Unipolar Converter for Switched Reluctance Motor Drives with Power Factor Improvement. In: APEC 2001. Sixteenth Annual IEEE Applied Power Electronics Conference and Exposition; 4–8 March; Anaheim, CA, USA: IEEE; 2001. p. 1103–1108. DOI: 10.1109/APEC.2001.912504

[19] Kwon YA, Shin JK, Rim GH. SRM Drive System with Improved Power Factor. In: Industrial Electronics, Control and Instrumentation, 1997. IECON 97. 23rd International Conference on; 14 Nov.; New Orleans, LA, USA: IEEE; 1997. p. 541–545. DOI: 10.1109/IECON.1997.671791

[20] Chai JY, Liaw CM. Development of a switched-reluctance motor drive with PFC front end. IEEE Transactions on Energy Conversion. 2009;**24**(1):30–42. DOI: 10.1109/TEC.2008.2002328

[21] Reinert J, Schroder S. Power-factor correction for switched reluctance drives. IEEE Transactions on Industrial Electronics. 2002;**49**(1):54–57. DOI: 10.1109/41.982248

[22] Goyal S, Kumar R, Gupta RA. Simulation and Analysis of Current Controlled PFC Converter-Inverter Fed SRM Drive. In: 2005 IEEE International Conference on Industrial Technology; 14–17 Dec; Hong Kong, China: IEEE; 2005. p. 1433–1437. DOI: 10.1109/ICIT.2005.1600860

[23] Kumar R, Gupta RA, Goyal S, Bishnoi SK. Fuzzy tuned PID controller based PFC converter-inverter fed SRM drive. In: 2006 IEEE International Conference on Industrial Technology; 2006. pp. 2498–2503. DOI: 10.1109/ICIT.2006.372632

[24] Rajesh M, Singh B. Power quality improvement in switched reluctance motor drive using Vienna rectifier. In: 2012 IEEE Fifth Power India Conference; 19–22 Dec; Murthal, India: IEEE; 2012. p. 1–7. DOI: 10.1109/PowerI.2012.6479580

[25] Rajesh M, Singh B. Analysis, design and control of single-phase three-level power factor correction rectifier fed switched reluctance motor drive. IET Power Electronics. 2014;**7**(6):1499–1508. DOI: 10.1049/iet-pel.2013.0621

[26] Chen YY, Hsu CT, Hu KW. On the PFC AC-DC Converter Fed SRM drives with Reversible and Regenerative Braking Capabilities. In: 2015 IEEE 2nd International Future Energy Electronics Conference (IFEEC); 1–4 Nov; Taipei, Taiwan: IEEE; 2015. p. 1–7. DOI: 10.1109/IFEEC.2015.7361545

[27] Singh B, Anand A. Improved Power Quality Zeta Converter Fed Switch Reluctance Motor Drive. In: 2015 2nd International Conference on Recent Advances in Engineering & Computational Sciences (RAECS); 21–22 Dec; Chandigarh, India: IEEE; 2015. p. 1–6. DOI: 10.1109/RAECS.2015.7453360

[28] Singh B, Anand A. A PFC Based SRM Motor Drive Using a Canonical Switching Cell Converter. In: 2016 IEEE 6th International Conference on Power Systems (ICPS); 4–6 March; New Delhi, India: IEEE; 2016. p. 1–6. DOI: 10.1109/ICPES.2016.7584246

[29] Chau KT. Switched Reluctance Motor Drives. In: Chau KT, editor. Electric Vehicle Machines and Drives: Design, Analysis and Application. 1st ed. eBook: Wiley-IEEE Press; 2015. p. 108–146. DOI: 10.1002/9781118752555.ch5

[30] Sehirli E, Kuyumcu FE, Cakir B. Current Controlled SRM fed by Three-phase Voltage Source Converter. In: 2015 50th International Universities Power Engineering Conference (UPEC); 1–4 Sept; Stoke on Trent, UK: IEEE; 2015. p. 1–6. DOI: 10.1109/UPEC.2015.7339858

[31] Singh B, Singh BN, Chandra A, Al-Haddad K, Pandey A, Kothari DP. A review of three-phase improved power quality AC-DC converters. IEEE Transactions on Industrial Electronics. 2004;**51**(3):641–660. DOI: 10.1109/TIE.2004.825341

[32] Sehirli E, Altınay M. Input-output linearization control of single-phase buck-boost power factor corrector. In: 2012 47th International Universities Power Engineering Conference (UPEC); 4–7 Sept; London, UK: IEEE; 2012. p. 1–6. DOI: 10.1109/UPEC.2012.6398557

[33] Slotine JJE, Li W. Applied Nonlinear Control. New Jersey: Prentice Hall; 1991

[34] Khalil HK. Nonlinear Systems. 3rd ed. New Jersey: Prentice Hall; 2002

[35] Lee TS. Input-output linearization and zero-dynamics control of three-phase AC/DC voltage-source converters. IEEE Transactions on Power Electronics. 2003;**18**(1):11–22. DOI: 10.1109/TPEL.2002.807145

[36] Kim CC, Hur J, Hyun DS. Simulation of a switched reluctance motors using Matlab/M-file. In: IEEE 2002 28th Annual Conference of the Industrial Electronics Society. IECON 02; 5–8 Nov; Sevilla, Spain: IEEE; 2002. p. 1066–1071. DOI: 10.1109/IECON.2002.1185420

[37] Huy LH/Mathworks. Switched Reluctance Motor [Internet]. Available from: https://www.mathworks.com/examples/simpower/mw/sps_product-power_SwitchedReluctanceMotor-switched-reluctance-motor [Accessed: March 2, 2017]

Chapter 6

Switched Reluctance Motor Drives for Hybrid Electric Vehicles

Christopher H.T. Lee, James L. Kirtley, Jr. and
M. Angle

Additional information is available at the end of the chapter

http://dx.doi.org/10.5772/intechopen.68911

Abstract

Because of the ever-increasing concerns on the energy utilization and environmental protection, the development of hybrid electric vehicles (HEVs) has become a hot research topic. As the major part of HEV technologies, the electric motor drives have to offer high efficiency, high power density, high controllability, wide-speed operating range, and maintenance-free operation. In particular, the switched reluctance (SR) motor drive can achieve most of these goals; therefore, this motor type has drawn much attention in the past. This chapter aims to serve as an overview of the latest developments of the SR motor drive, purposely for HEV applications. To be specific, the discussions on motor structures for torque density enhancement and torque ripple minimization are covered.

Keywords: electric vehicle, hybrid electric vehicle, motor drive, switched reluctance, torque density enhancement, torque ripple minimization

1. Introduction

With ever-enhancing concerns on energy crisis and environmental pollution, as one of the most promising solutions, the development of electric vehicles (EVs) has been speeding up in recent years [1, 2]. However, the traditional EVs are powered by battery as the sole energy source, so that this particular EV type can only provide support for short-duration trips and is not favourable for typical daily drivers [3, 4]. To extend the normal driving range and to relieve the problem, the hybrid electric vehicle (HEV), which incorporates an additional internal combustion engine (ICE) as the supplementary source, has drawn extensive attention in return [5].

As the major component for the HEV systems, the electric motor drives have to fulfil several criteria [6, 7], including

- high efficiency,
- high power and torque densities,
- high controllability,
- wide-speed operating range,
- maintenance-free operation,
- high reliability and robustness,
- high cost-effectiveness.

Since the interior permanent-magnet (IPM) motor drives can achieve most of the desired goals, these types of motor drives have been widely adopted in some famous HEV systems, such as Toyota Prius HEV [8]. Even though the IPM motor drive can offer very desirable power and torque performances for HEV applications, it suffers from the problem of high PM material costs and uncontrollable PM flux densities [9, 10]. Hence, the magnetless switched reluctance (SR) motor drives, which offer satisfactory cost-effectiveness and desirable flux controllability, have become more popular recently [11, 12].

The purpose of this chapter is to serve as an overview of the latest SR motor developments in HEV applications. The current and upcoming design philosophies of the SR motor drives will be discussed. In addition, quantitative comparisons among the latest designs will also be covered.

2. Hybrid electric vehicle architecture

The fundamental HEV can be defined as the vehicle system that consists of two or more power sources, while there are numerous combinations available in the domestic market. Nowadays, the most common HEV architecture includes the electric motor drive and the ICE. In general, the typical HEV system can be classified into two major groups, namely, the series HEV and the parallel HEV, as shown in **Figures 1** and **2**, respectively.

In the series HEV system, the HEV is powered solely by the electric motor drive, while the motor drive can be supplied either from the battery or from ICE-driven generator unit, or even both [3]. On the other hand, in the parallel HEV system, the HEV can be powered by either the electric motor drives or ICE, or by both simultaneously.

In both the series and parallel HEV systems, the battery can be recharged by the ICE during normal operation, so that the battery capacity as well as the battery size can be greatly reduced, as compared with the EV systems. Similar to the EV system, the HEV systems can also offer the regenerative braking operation, such that the battery can be recharged by the electric motor drives when the HEV undergoes braking conditions. As a result, the overall fuel efficiency can be greatly improved, as compared with the standard ICE vehicles.

Since the series HEV relies solely on the motor drive to power the HEV itself, the motor drive is relatively larger and more expensive, as compared with those employed in the parallel HEV

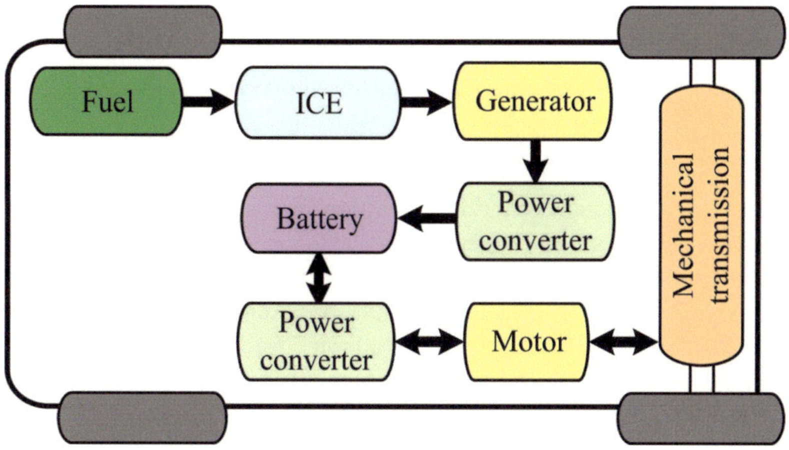

Figure 1. Series hybrid electric vehicle system.

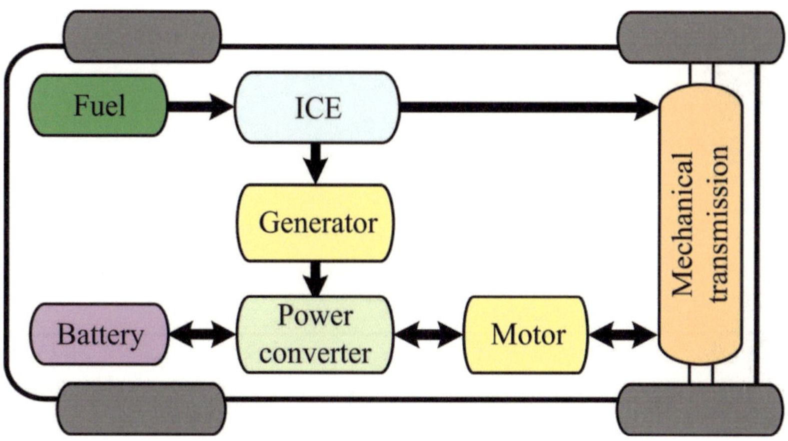

Figure 2. Parallel hybrid electric vehicle system.

counterparts. On the other hand, the parallel HEV needs not to be powered barely from electric motor drives, so that the motor sizing can be reduced. By taking all the criteria into consideration, the series HEV is generally favourable for specialist applications, while the parallel HEV for broad applications.

3. Switched reluctance motor drives

With the definite advantages of low material cost, robust structure, mature converter topology, high efficiency, satisfactory power and torque densities, and simple control algorithm, the SR motor drives have been considered as a promising candidate for HEV applications. Without any installation of PM material or winding on its rotor, the SR motor drives enjoy the higher cost-effectiveness and wide-speed operation range, as compared with its counterparts. To be specific, unlike the induction and PM motor drives, SR motor drives can relieve the mechanical problems caused by the centripetal forces at high-speed operation.

The SR motor drives consist of the simple structure, while its design philosophies require sophisticated analysis and researches. Because of the inherited doubly salient topology, the SR motor drives result in nonlinear inductance value, high saturation in pole tips, and severe fringing effect among the poles. All these characteristics increase the design difficulty when the magnetic circuit approach is employed. In general, the finite element method (FEM) is known as the most accurate and convenient tool to analyse the SR motor drives [7]. In this chapter, the commercial FEM software, JMAG-Designer, is adopted for the motor performance analysis. Upon the employment of the FEM-based electromagnetic analysis, the key parameters and performances of the SR motor drives can be provided.

3.1. Stator and rotor pole arrangement

To operate the SR motor drives properly, its stator and rotor poles must obey certain specific regulations. To be specific, the stator and rotor poles should be equally distributed among the circumferences, so that the pole arrangement is governed as

$$\begin{cases} N_s = 2mk \\ N_r = N_s \pm 2k \end{cases} \tag{1}$$

where N_s is the number of stator poles, N_r is the number of rotor poles, m is the number of phases and k is a positive integer. The fundamental pole combination of the SR motor drives is shown in **Table 1**.

To minimize the switching frequency as well as the core losses, the stator poles are generally chosen to be larger than the rotor poles, i.e. $N_s > N_r$ in Eq. (1). A typical example of a three-phase 6/4-pole SR motor drive is shown in **Figure 3**. By taking the reliability and cost-effectiveness into consideration, the three-phase and four-phase SR motor drives are the equally common candidates employed in the domestic markets. The three-phase SR motor drive takes the definite advantage of a larger winding slot area, as compared with the four-phase counterpart. However, with the inherited larger dead zones, the three-phase SR motor drive suffers

m	k	N_s	N_r	N_r
3	1	6	8	4
4	1	8	10	6
5	1	10	12	8
3	2	12	16	8
4	2	16	20	12

Table 1. Fundamental pole combinations of switched reluctance motor drives.

http://dx.doi.org/10.5772/intechopen.68911

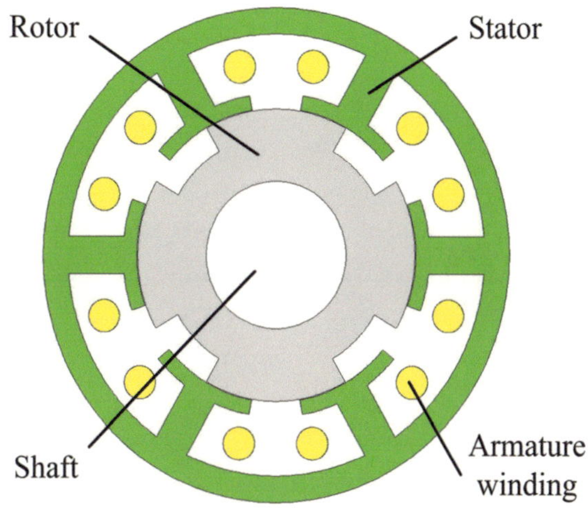

Figure 3. Three-phase 6/4-pole switched reluctance motor.

from a larger torque ripple problem. The torque pulsation can be relieved based on the larger number of poles per phase structure, i.e. the three-phase 12/8-pole structure.

3.2. Operating principle

Since the SR motor drives consist of the doubly salient structure, its reluctance of the magnetic flux path varies along the stator-rotor position. The torque can then be produced based on the 'minimum reluctance' rule, i.e. the rotor pole has the tendency to align with the excited stator pole such that the reluctance of the magnetic flux can be minimized.

The operating principle and the theoretical waveform are shown in **Figure 4**. To produce a positive electromagnetic torque T, the armature current i should be injected when the self-inductance L is increasing. The torque production can be described as

$$T = \frac{1}{2} i^2 \frac{dL}{d\theta} \tag{2}$$

The operating speed n of the SR motor drives is governed by the number of rotor poles N_r and the operating frequency f as

$$n = \frac{60f}{N_r} \tag{3}$$

In general, the larger the number of rotor poles, the lower the operating speed in the given operating frequency. Therefore, the SR motor drives with a larger number of rotor poles, e.g. the three-phase 12/8-pole and the four-phase 16/12-pole structures, are more preferable for low-speed applications, such as the direct-drive HEV system.

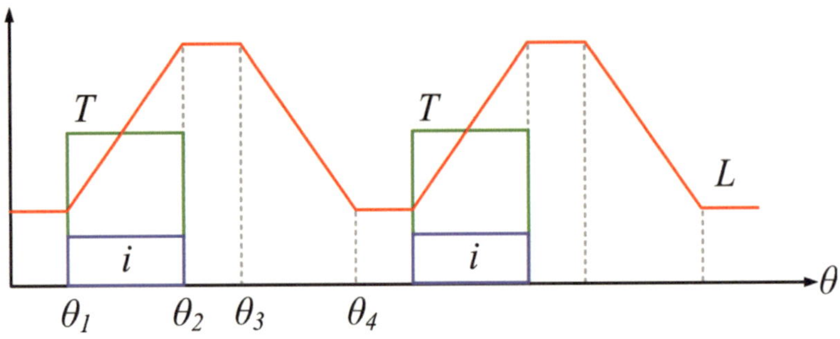

Figure 4. Theoretical operating waveform.

4. Torque density enhancement

4.1. Double-stator structure

As one of the most important criteria for HEV applications, the enhancement of torque density for SR motor drives has become a hot research area. Similar to other commonly employed motor drives, the conventional SR motor drives generally adopt the outer-stator and inner-rotor structure in most occasions. With this traditional topology, the inner spacing of the motor drives is heavily wasted. To improve the situation and to utilize the inner spacing for torque production, the double-stator (DS) topology has been developed [13].

In general cases, the outer and inner segments of the DS motor drives are duplicated, so that both of them employ the same pole arrangement. Since the outer and inner segments are symmetric, the outer and inner armature windings can be connected in series to simplify the operating algorithm. An example of the three-phase 6/4-pole DS-SR motor drive is shown in **Figure 5**. Upon this design, both the outer and inner stators can transfer power to the rotor simultaneously, so that the torque density of the DR motor drives can be greatly improved, as compared with the single-stator counterparts. Because of the additional stator construction, the DS motor drives suffer from higher manufacturing complexity. Yet, the SR motor drives are free from PM material installations, so that the corresponding manufacturing processes are still acceptable and feasible in practical situations.

4.2. Performance analysis of double-stator structure

Based on Eq. (1), the three-phase 12/8-pole single-stator SR motor drive and the three-phase 12/8-pole DS-SR motor drive are designed and are shown in **Figures 6** and **7**, respectively. The three-phase single-stator SR motor consists of the outer-stator and inner-rotor structure, while the three-phase DS-SR motor consists of the dual-stator sandwiched rotor structure. The outer and inner segments of the DS-SR motor are purposely designed with the symmetrical pattern, such that the armature coils can be connected in series. The proposed SR motor drives are designed based on the specification of the typical HEV applications [14], while the key design data are listed in **Table 2**.

http://dx.doi.org/10.5772/intechopen.68911

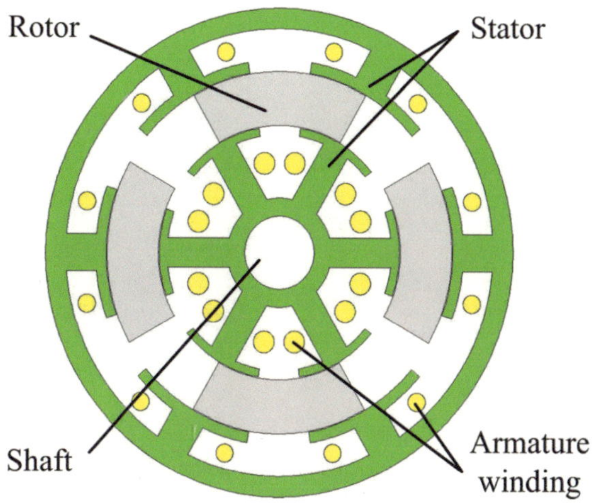

Figure 5. Three-phase 6/4-pole double-stator switched reluctance motor.

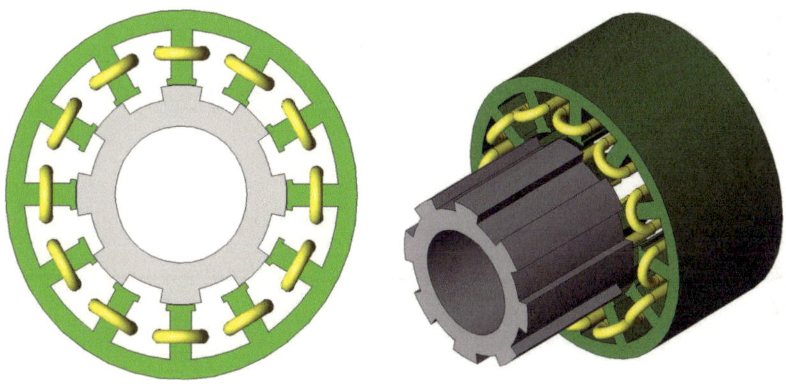

Figure 6. Three-phase 12/8-pole single-stator switched reluctance motor.

To offer a fair comparison, all the key parameters, namely the outer-stator outside diameter, airgap length, stack length, and slot fill factor are set equal. In the meantime, the motor drives are purposely optimized to avoid the magnetic saturation, such that the corresponding core losses can be minimized.

The magnetic field distributions of the SR motor and the DS-SR motor are shown in **Figures 8** and **9**, respectively. As shown, the field distribution of the two proposed motors is well balanced with no obvious saturations. The results somehow give the evidence that the proposed motors are optimized after the iterative approach, in a way their power losses can be minimized. Since the DS-SR motor consists of a symmetrical structure between the outer and the inner segments, the magnetic field distributions between the two segments are also symmetrical.

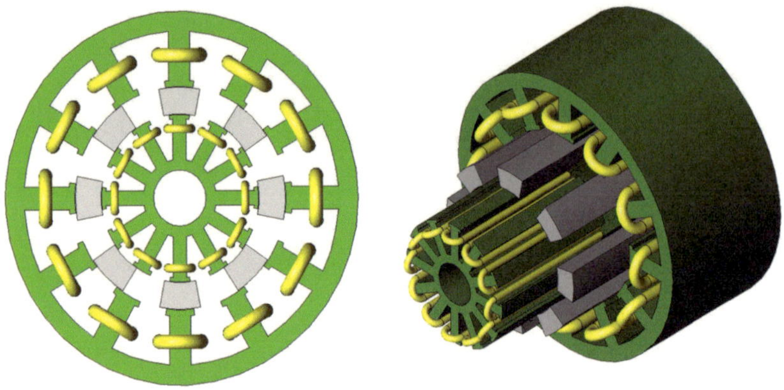

Figure 7. Three-phase 12/8-pole double-stator switched reluctance motor.

Item	SR	DS-SR	MO-DS-SR
Number of stator pole	12	12	12
Number of rotor pole	8	8	8
Outside-stator outside diameter	269 mm	269 mm	269 mm
Outside-stator inside diameter	161 mm	161 mm	161 mm
Rotor outside diameter	160 mm	160 mm	160 mm
Rotor inside diameter	100 mm	121 mm	121 mm
Inside-stator outside diameter	N/A	120 mm	120 mm
Inside-stator inside diameter	N/A	40 mm	40 mm
Airgap length	0.5 mm	0.5 mm	0.5 mm
Stack length	135 mm	135 mm	135 mm
Slot fill factor	55%	55%	55%
Number of phase	3	3	3
Number of outer armature turn	50	50	50
Number of inner armature turn	N/A	18	18
Mechanical offset	0°	0°	60°

Table 2. Key design data of proposed motor drives.

The airgap flux density of the SR motor, the outer airgap density and the inner-airgap density of the DS-SR motor are shown in **Figures 10–12**, respectively. Since the outer and inner segments of the DS-SR motor are symmetrical, as expected, its outer and inner airgaps are symmetrical in nature. In the meantime, not surprisingly, the single stator and DS-SR motor share the same flux patterns because both of them share the same motor structure and operating principle.

http://dx.doi.org/10.5772/intechopen.68911

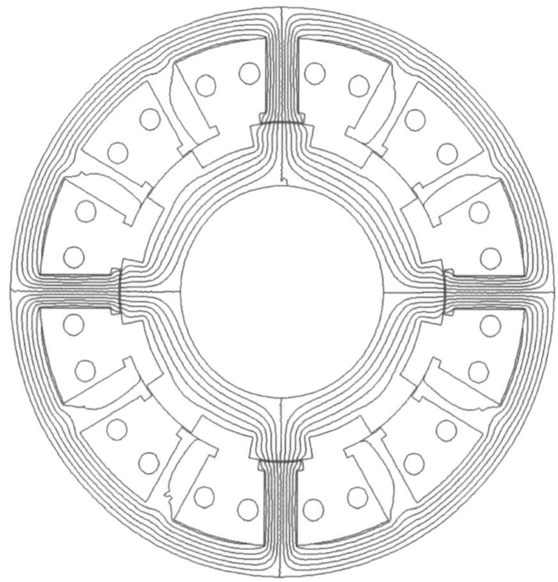

Figure 8. Magnetic field distribution of the switched reluctance motor.

Figure 9. Magnetic field distribution of the double-stator switched reluctance motor.

The torque capabilities of the SR motor and the DS-SR motor are given in **Figures 13** and **14**, respectively. The motor drives are conducted with various armature currents from 50 to 200 Å with a step of 50 Å. The torque capabilities of both motors follow the description from Eq. (2),

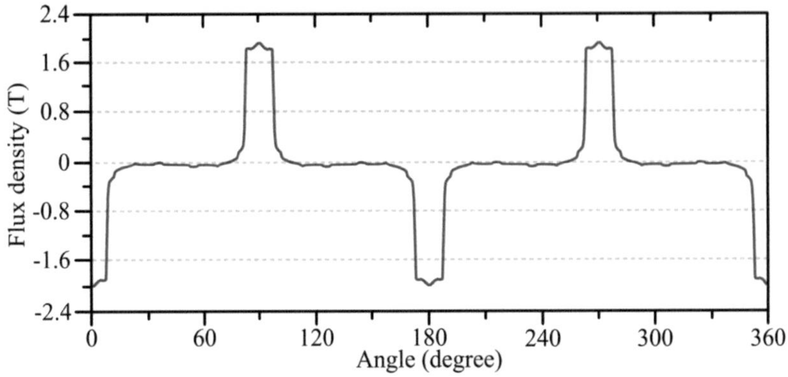

Figure 10. Airgap flux density of the switched reluctance motor.

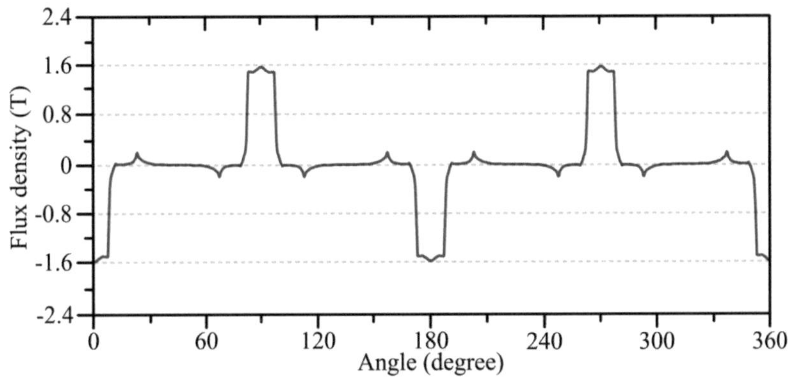

Figure 11. Outer-airgap flux density of the double-stator switched reluctance motor.

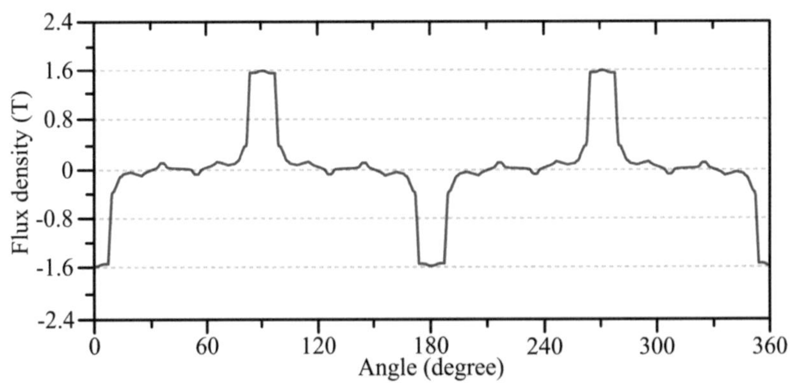

Figure 12. Inner-airgap flux density of the double-stator switched reluctance motor.

http://dx.doi.org/10.5772/intechopen.68911

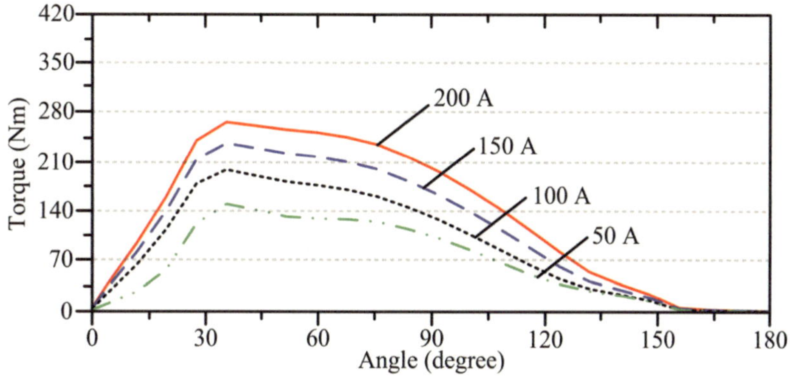

Figure 13. Torque capabilities of the switched reluctance motor.

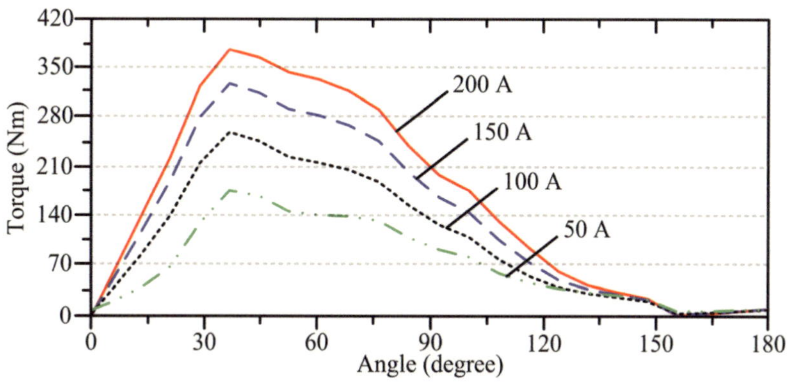

Figure 14. Torque capabilities of the double-stator switched reluctance motor.

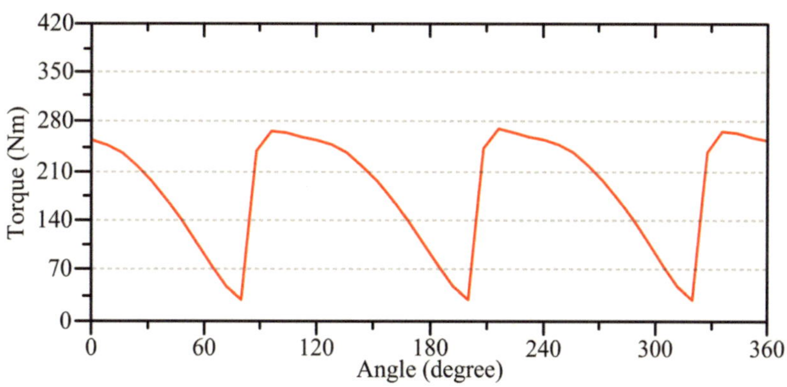

Figure 15. Steady torque waveform of the switched reluctance motor.

such that it can be confirmed that the proposed motors consist of a minimized magnetic saturation problem. In addition, it can also be confirmed that with the DS structure, the DS-SR motor can produce higher torque capability than its single-stator counterpart.

The steady output torques of the SR and the DS-SR motors are given in **Figures 15** and **16**, respectively. As shown, the average torque of the SR motor is about 178 Nm, while of the DS-SR motor is 242 Nm. The torque enhancement from the DS structure can be up to 36%. The performance comparisons between the two proposed motors, namely, the single-stator SR motor and the DS-SR motor are summarized in **Table 3**.

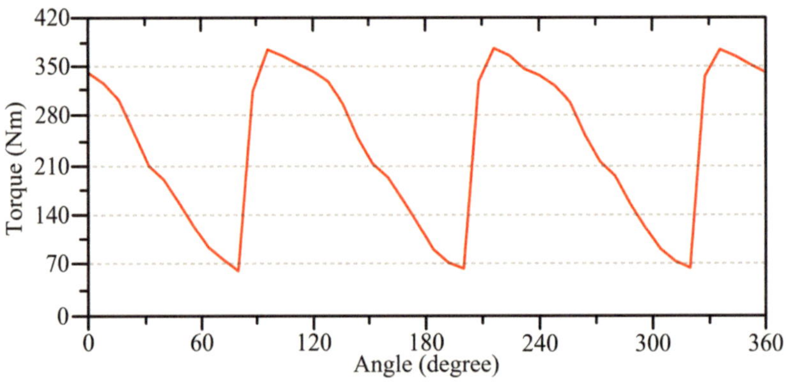

Figure 16. Steady torque waveform of the double-stator switched reluctance motor.

Item	SR	DS-SR
Rated power	22 kW	30 kW
Base speed	1200 rpm	1200 rpm
Outer-airgap flux density	1.9 T	1.6 T
Inner-airgap flux density	N/A	1.6 T
Rated torque	178 Nm	242 Nm
Torque enhancement	N/A	36%

Table 3. Performance comparisons between the torque density enhancement structures.

5. Torque ripple minimization

5.1. Mechanical-offset structure

The DS structures can definitely improve the torque densities, while the torque ripple value, another key criterion to determine the torque performance, should also be discussed. The torque vibration in the SR motor drives can be explained from the torque production processes, i.e. the

http://dx.doi.org/10.5772/intechopen.68911

torque pulsation generated by the commutation between different phases. To quantitatively analyse the torque ripple performance, the torque ripple factor K_T is widely employed as

$$K_T = \frac{T_{max} - T_{min}}{T_{avg}} \times 100\% \tag{4}$$

where T_{max}, T_{min}, and T_{avg} are the maximum, minimu and average torque produced, respectively.

According to Eq. (4), the torque pulsation value is inversely related to the average torque value, so that mathematically the torque ripple can be reduced upon the increase of average torque. As aforementioned, the conventional DS motor drives generally employ the symmetrical structure between the outer and inner segments, such that the outer and inner rotor teeth are aligned with each other. Upon this arrangement, the local maxima and local minima of the torque ripples from the outer and inner segments are unfavourably integrated, such that the increased torque level can barely benefit the torque ripple value.

On the other hand, the mechanical-offset (MO) structure, which purposely mismatches the outer and inner rotor teeth with a conjugated angle to each other, can greatly improve the torque pulsation problem [15]. Upon the MO structure, the local torque ripple maxima and local minima from the outer and inner segments are favourably offset with each other, such that the torque ripple problem can be relieved.

5.2. Conduction algorithm for conventional double-stator motor drive

As described in Section 3.2, to produce an electromagnetic torque, a unipolar rectangular armature current should be injected in accordance with the status of the self-inductance with conduction angle $\theta_c = \theta_2 - \theta_1$, as further elaborated in **Figure 17**. Since the conventional DS motor drive employs the symmetrical structure between its outer and inner segments, the armature windings in the outer and inner stators of the conventional DS motor drive are connected in series. Consequently, to operate the motor drive properly, the armature currents in both stators are identical as

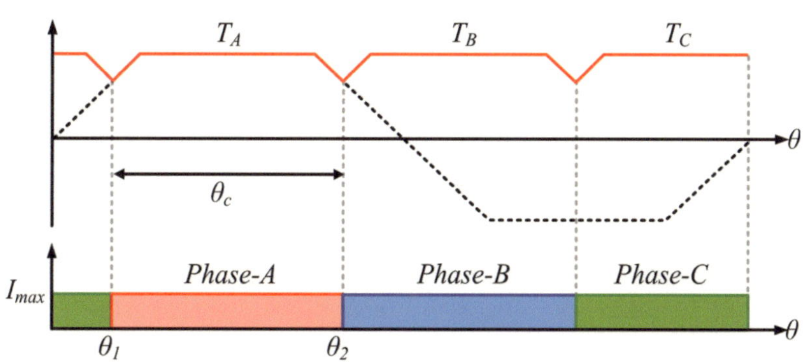

Figure 17. Operating principle for switched reluctance motor.

$$\begin{cases} i_k = I_{\max} \; \theta_1 \leq \theta \leq \theta_2 \\ i_k = 0 \; 0 < \theta < \theta_1, \; \theta_2 < \theta < 2\pi \end{cases} \text{for } k = 1, 2 \tag{5}$$

where i_1 and i_2 are the armature windings in the outer and inner stators, respectively. Using this conduction algorithm, as shown in **Figure 18**, the power from both stators can be transferred to the rotor simultaneously, so that the torque density of the DS motor can be boosted. However, the local maxima and local minima of the two individual torque components are unfavourably integrated with each other. As a result, the torque ripple value remains the same, as compared with the single-stator counterpart.

5.3. Conduction algorithm for mechanical-offset motor drive

As described in Eq. (2), the average torque magnitude is governed by the relative position between the stator and rotor teeth. Consequently, the local maxima and local minima of the torque ripple are also governed by the relative position between the stator and rotor teeth. Upon the MO structure, the outer and inner rotor teeth of the SR motor drive are artfully mismatched with a conjugated electrical angle of $\theta_m = \pi/m$, as shown in **Figure 19**. Since the outer and inner rotor teeth no longer align with each other, the armature currents in the two stators should be operated separately. To be specific, either one set of the armature winding should be injected with current described in Eq. (5), while the other set as follows

$$\begin{cases} i_{1\ or\ 2} = I_{\max} \; \theta_1 + \theta_m \leq \theta \leq \theta_2 + \theta_m \\ i_{1\ \text{or}\ 2} = 0 \; \theta_m < \theta < \theta_1 + \theta_m, \; \theta_2 + \theta_m < \theta < 2\pi + \theta_m \end{cases} \tag{6}$$

Upon this conduction scheme, the local maxima and local minima of the torque ripples are artfully offset with each other, so that the torque pulsations generated by the two torque components can be favourably compensated. As a result, the smoother resultant torque, as compared with those produced by the conventional DS motor drives, can be generated.

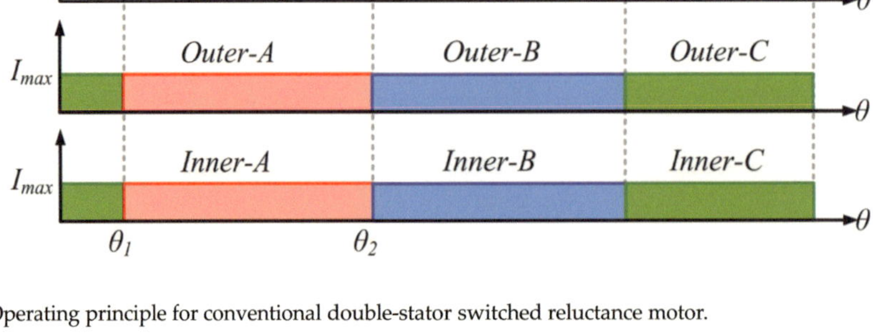

Figure 18. Operating principle for conventional double-stator switched reluctance motor.

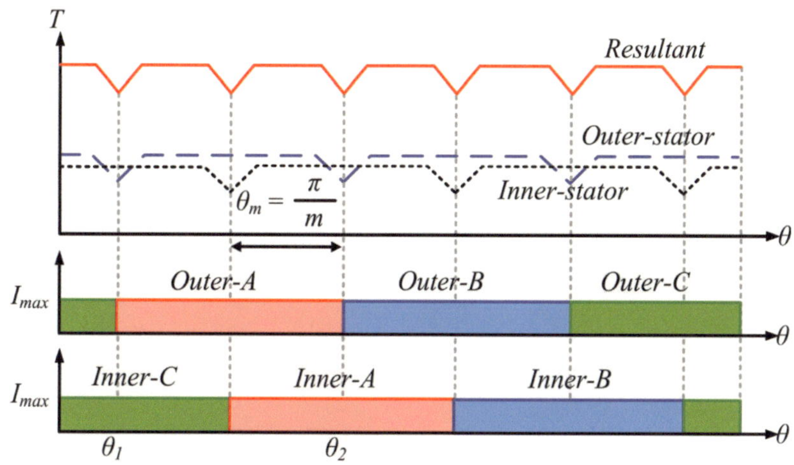

Figure 19. Operating principle for mechanical-offset double-stator switched reluctance motor.

5.4. Torque characteristics of the mechanical-offset motor drive

Since the two torque components of the MO motor drive are integrated with each other, in principle, its resultant torque level should not be affected. Therefore, the MO motor drive should be able to produce the same torque level, as compared with the conventional DS counterpart. To offer a better torque ripple minimization effect, the magnitudes between the two torque components should be adjusted to similar levels. This can be handled by modifying the design of the motor dimensions, i.e. the electric loadings between the outer and inner segments, which are out of scope of this chapter.

Unlike the DS motor drives that align the local torque ripple maxima and local minima together, the MO motor drive purposely mismatches these torque ripples. As a result, the local maxima and local minima are spread according to the conjugated position, such that the pulsating frequency is double, as compared with the DS counterpart.

5.5. Performance analysis of mechanical-offset structure

Based on Eq. (1), the three-phase 12/8-pole MO-DS-SR motor drive is designed and shown in **Figure 20**. To offer a fair comparison, the MO-DS-SR motor is purposely designed based on the same conditions in Section 4.2. Both the DS-SR motor and the MO-DS-SR motor share the same motor structure, while the major distinction comes from the rotor alignments, i.e. the outer and inner rotor teeth of the DS-SR motor align with each other, while those of the MO-DS-SR motor are purposely mismatched with a conjugated angle. The key design data of the MO-DS-SR motor are listed in **Table 2**.

The steady torques of the DS-SR motor and the MO-DS-SR motor are shown in **Figures 21** and **22**, respectively. The average rated torque of the outer-stator, the inner-stator, and both stators of the DS-SR motor are 178, 68, and 242 Nm, respectively, while those of the MO-DS-SR motor are 178, 68, and 239 Nm, respectively. It can be confirmed that the two torque components of the MO-DS-SR motor are seamlessly integrated with each other, so that the same torque level, as compared with the DS-SR counterpart, can be produced. With the MO structure, the local torque ripple maxima and local minima are favourably offset with each other, so that the MO-DS-SR motor can reduce its

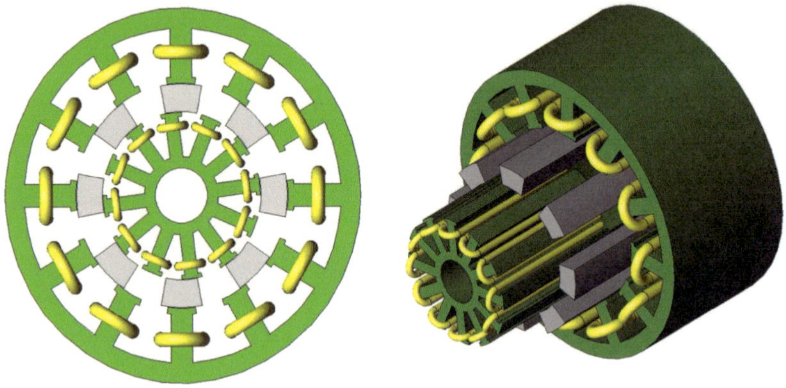

Figure 20. Three-phase 12/8-pole mechanical-offset double-stator switched reluctance motor.

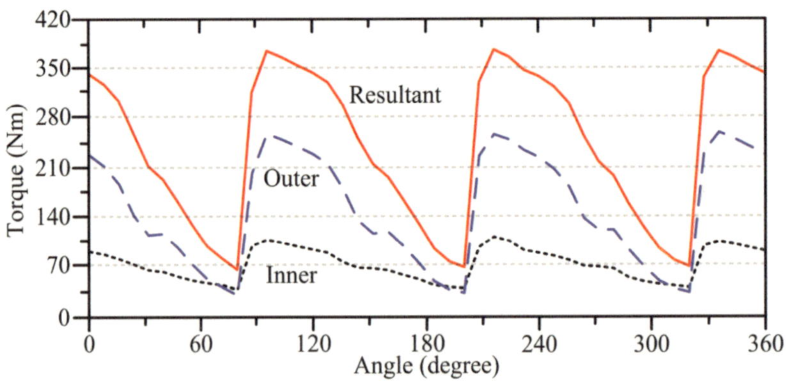

Figure 21. Steady torque waveforms of the conventional double-stator switched reluctance motor.

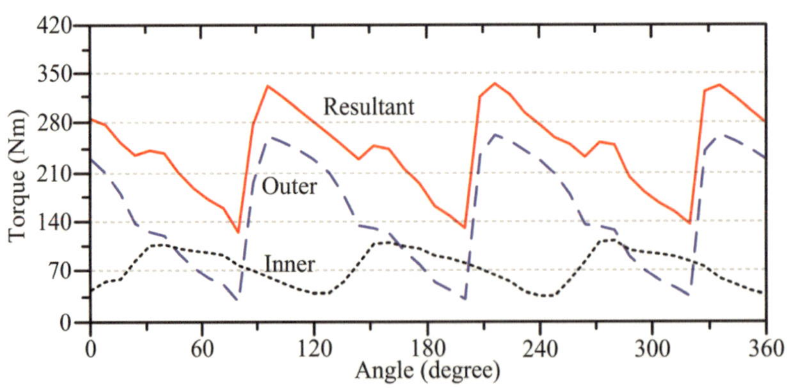

Figure 22. Steady torque waveforms of the mechanical-offset double-stator switched reluctance motor.

Item	DS-SR	MO-DS-SR
Rated power	30 kW	30 kW
Base speed	1200 rpm	1200 rpm
Outer-stator torque	178 Nm	178 Nm
Inner-stator-torque	68 Nm	68 Nm
Resultant torque	242 Nm	239 Nm
Torque ripple	124%	81%

Table 4. Performance comparisons between the torque ripple minimization structures.

torque ripple value to 81%, as compared with those of 124% in the DS-SR motor. The performance comparisons among the two proposed motors, namely, the DS-SR motor and the MO-DS-SR motor are summarized in **Table 4**.

6. Conclusion

In this chapter, the design criteria, principles of operation, and the key performances of the SR motor drives are carefully discussed and quantitatively compared, based on the fair conditions. To be specific, the torque density enhancement structure, namely the DS structure, and the torque ripple minimization structure, namely the MO structure, are included. This chapter provides the latest research trend on SR motor drives for HEV applications, such that it can serve as the blueprint and start-up manual for the potential readers to develop the research interests in this particular area.

Author details

Christopher H.T. Lee*, James L. Kirtley, Jr. and M. Angle

*Address all correspondence to: chtlee@mit.edu

Research Laboratory of Electronics, Massachusetts Institute of Technology, Cambridge, Massachusetts, United States

References

[1] Chan CC. The state of the art of electric, hybrid, and fuel cell vehicles. Proceedings of the IEEE. 2007;**95**(4):704–718

[2] Emadi A, Lee YJ, Rajashekara K. Power electronics and motor drives in electric, hybrid electric, and plug-in hybrid electric vehicles. IEEE Transactions on Industrial Electronics. 2008;**55**(6):2237–2245

[3] Ehsani M, Gao Y, Gay SE, Emadi A. Modern Electric, Hybrid Electric, and Fuel Cell Vehicle. Boca Raton, FL: CRC Press; 2004

[4] Zhu ZQ, Howe D. Electrical machines and drives for electric, hybrid, and fuel cell vehicles. Proceedings of the IEEE. 2007;**95**(4):746–765

[5] Lee CHT, Liu C, Chau KT. A magnetless axial-flux machine for range-extended electric vehicles. Energies. 2014;**7**(3):1483–1499

[6] Cheng M, Hua W, Zhang J, Zhao W. Overview of stator-permanent magnet brushless machines. IEEE Transactions on Industrial Electronics. 2011;**58**(11):5087–5101.

[7] Chau KT. Electric Vehicle Machines and Drives: Design, Analysis and Application. Singapore:Wiley-IEEE Press; 2015

[8] Cao R, Mi C, Cheng M. Quantitative comparison of flux-switching permanent-magnet motors with Interior permanent magnet motor for EV, HEV, and PHEV applications. IEEE Transactions on Magnetics. 2008;**48**(8):2274–2384

[9] Boldea I, Tutelea LN, Parsa L, Dorrell D. Automotive electric propulsion systems with reduced or no permanent magnets: An overview. IEEE Transactions on Industrial. Electronics. 2014;**61**(10):5696–5711

[10] Lee CHT, Chau KT, Liu C. Design and analysis of an electronic-geared magnetless machine for electric vehicles. IEEE Transactions on Industrial. Electronics. 2016;**63**(11):6705–6714

[11] Rahman KM, Fahimi B, Suresh G, Rajarathnam AV, Ehsani E. Advantages of switched reluctance motor applications to EV and HEV: Design and control issues. IEEE Transactions on Industry Applications. 2000;**36**(1):111–121

[12] Chiba A, Takano Y, Takeno M, Imakawa T, Hoshi N, Takemoto M, Ogasawara S. Torque density and efficiency improvements of a switched reluctance motor without rare-earth switched reluctance motor without rare-earth. IEEE Transactions on Industry Applications. 2011;**47**(3):1240–1246

[13] Abbasian M, Moallem M, Fahimi B. Double-stator switched reluctance machines (DSSRM): Fundamentals and magnetic force analysis. IEEE Transactions on Energy Conversion. 2010;**25**(3):589–597

[14] Takeno M, Chiba A, Hoshi N, Ogasawara S, Takemoto M, Rahman MA. Test results and torque improvement of the 50-kW switched reluctance motor designed for hybrid electric vehicles. IEEE Transaction on Industry Applications. 2012;**48**(4):1327–1334

[15] Lee CHT, Chau KT, Liu C, Ching TW, Li F. Mechanical offset for torque ripple reduction for magnetless double-stator doubly salient machine. IEEE Transactions on Magnetics. 2014;**50**(11):8103304